T0321992

Information Security Practices for the Internet of Things, 5G, and Next–Generation Wireless Networks

Biswa Mohan Sahoo
Manipal University, Jaipur, India

Suman Avdhesh Yadav
Amity University, India

A volume in the Advances in
Information Security, Privacy, and
Ethics (AISPE) Book Series

Published in the United States of America by
    IGI Global
    Information Science Reference (an imprint of IGI Global)
    701 E. Chocolate Avenue
    Hershey PA, USA 17033
    Tel: 717-533-8845
    Fax: 717-533-8661
    E-mail: cust@igi-global.com
    Web site: http://www.igi-global.com

Library of Congress Cataloging-in-Publication Data

Names: Sahoo, Biswa, 1982- editor. | Yadav, Suman, 1986- editor.
Title: Information security practices for the Internet of Things, 5G, and
  next-generation wireless networks / Biswa Sahoo and Suman Yadav,
  editors.
Description: Hershey, PA : Information Science Reference, an imprint of IGI
  Global, [2022] | Includes bibliographical references and index. |
  Summary: "This book highlights research on secure communication of 5G,
  Internet of Things (IoT) and Next-Generation wireless networks, along
  with related areas to ensure secure and Internet-compatible IoT
  systems"-- Provided by publisher.
Identifiers: LCCN 2022970056 (print) | LCCN 2022970057 (ebook) | ISBN
  9781668439210 (h/c) | ISBN 9781668439227 (s/c) | ISBN 9781668439234
  (ebook)
Subjects: LCSH: Internet of things--Security measures. | 5G mobile
  communication systems--Security measures.
Classification: LCC TK5105.8857 .I4854 2022 (print) | LCC TK5105.8857
  (ebook) | DDC 004.67/8--dc23/eng/20220304
LC record available at https://lccn.loc.gov/2022970056
LC ebook record available at https://lccn.loc.gov/2022970057

This book is published in the IGI Global book series Advances in Information Security, Privacy,
and Ethics (AISPE) (ISSN: 1948-9730; eISSN: 1948-9749)

British Cataloguing in Publication Data
A Cataloguing in Publication record for this book is available from the British Library.

All work contributed to this book is new, previously-unpublished material.
The views expressed in this book are those of the authors, but not necessarily of the publisher.

For electronic access to this publication, please contact: eresources@igi-global.com.

Advances in Information Security, Privacy, and Ethics (AISPE) Book Series

ISSN:1948-9730
EISSN:1948-9749

Editor-in-Chief: Manish Gupta State University of New York, USA

MISSION

As digital technologies become more pervasive in everyday life and the Internet is utilized in ever increasing ways by both private and public entities, concern over digital threats becomes more prevalent.

The **Advances in Information Security, Privacy, & Ethics (AISPE) Book Series** provides cutting-edge research on the protection and misuse of information and technology across various industries and settings. Comprised of scholarly research on topics such as identity management, cryptography, system security, authentication, and data protection, this book series is ideal for reference by IT professionals, academicians, and upper-level students.

COVERAGE

- Computer ethics
- Security Information Management
- Privacy-Enhancing Technologies
- Cyberethics
- Global Privacy Concerns
- IT Risk
- Information Security Standards
- Security Classifications
- Network Security Services
- Electronic Mail Security

IGI Global is currently accepting manuscripts for publication within this series. To submit a proposal for a volume in this series, please contact our Acquisition Editors at Acquisitions@igi-global.com or visit: http://www.igi-global.com/publish/.

Titles in this Series

For a list of additional titles in this series, please visit:
www.igi-global.com/book-series/advances-information-security-privacy-ethics/37157

Global Perspectives on Information Security Regulations Compliance, Controls, and Assurance
Guillermo A. Francia III (University of West Florida, USA) and Jeffrey S. Zanzig (Jacksonville State University, USA)
Information Science Reference • © 2022 • 309pp • H/C (ISBN: 9781799883906) • US $240.00

Handbook of Research on Cyber Law, Data Protection, and Privacy
Nisha Dhanraj Dewani (Maharaja Agrasen Institute of Management Studies, Guru Gobind Singh Indraprastha University, India) Zubair Ahmed Khan (University School of Law and Legal Studies, Guru Gobind Singh Indraprastha University, India) Aarushi Agarwal (Maharaja Agrasen Institute of Management Studies, India) Mamta Sharma (Gautam Buddha University, India) and Shaharyar Asaf Khan (Manav Rachna University, India)
Information Science Reference • © 2022 • 390pp • H/C (ISBN: 9781799886419) • US $305.00

Cybersecurity Crisis Management and Lessons Learned From the COVID-19 Pandemic
Ryma Abassi (Sup'Com, University of Carthage, Tunisia) and Aida Ben Chehida Douss (Sup'Com, University of Carthage, Tunisia)
Information Science Reference • © 2022 • 276pp • H/C (ISBN: 9781799891642) • US $240.00

Applications of Machine Learning and Deep Learning for Privacy and Cybersecurity
Anacleto Correia (CINAV, Portuguese Naval Academy, Portugal) and Victor Lobo (Nova-IMS, Naval Academy, Portugal)
Information Science Reference • © 2022 • 315pp • H/C (ISBN: 9781799894308) • US $250.00

Cybersecurity Capabilities in Developing Nations and Its Impact on Global Security
Maurice Dawson (Illinois Institute of Technology, USA) Oteng Tabona (Botswana International University of Science and Technology, Botswana) and Thabiso Maupong (Botswana International University of Science and Technology, Botswana)

701 East Chocolate Avenue, Hershey, PA 17033, USA
Tel: 717-533-8845 x100 • Fax: 717-533-8661
E-Mail: cust@igi-global.com • www.igi-global.com

Table of Contents

Chapter 12
Scheduling Optimization Based on Energy Prediction Using ARIMA Model
Pooja Chaturvedi, Institute of Technology, Nirma University,
Ahmedabad, India
Ajai Kumar Daniel, Madan Mohan Malaviya University of Technology,
Gorakhpur, India

Detailed Table of Contents

Chapter 1
Sridevi, Karnatak University, Dharwad, India
Tukkappa K. Gundoor, Karnatak University, Dharwad, India

Internet of things (IoT) meets 5G communications, which aims to use a variety of promising network technologies to support a significant number of connected devices. For cognitive computing, massive machine type of communication (mMTC), cloud computing, artificial intelligence (AI), and so on, huge challenges for security, privacy, and trust are predicted. Technologies for 5G wireless communication encourage its use of mobile networks not just to connect with people and machines but also to connect and manage other equipment that supports virtual reality, such as self-driving cars, IoT drones, surveillance, and security. It is also critical to safeguard the technology of the 5G networks for IoT communication from threats. The different models for 5G IoT communication environment, vulnerabilities, attacks, and its several security protocols are described. The current security mechanisms in 5G networks IoT nature were analyzed and compared. The security challenges and future orientations of 5G-based systems are discussed.

Chapter 2
Sumit Dhariwal, Manipal University Jaipur, Jaipur, India
Avani Sharma, Manipal University Jaipur, Jaipur, India

5G could help with the extremely dependable and cost-effective networking of a huge number of devices (e.g., internet of things [IoT]), as well as universal broadband access and high user mobility. The current technical enablers for 5G are cloud computing, software-defined networking (SDN), and service-oriented virtualization (NFV). However, these technologies offer security issues in addition to raising concerns about user privacy. In this chapter, the authors give an outline of the security difficulties that these technologies face, as well as the privacy rules that apply to 5G. They also offer security remedies to these problems, as well as research directions for dependable 5G frameworks.

Chapter 3

Anita Patil, S. G. Balekundri Institute of Technology, India
Sridhar Iyer, KLE Dr. M. S. Sheshgiri College of Engineering and
Technology, India
Rahul J. Pandya, Indian Institute of Technology, Dharwad, India

Over the past decade, in view of minimizing network expenditures, optimizing network performance, and building new revenue streams, wireless technology has been integrated with artificial intelligence/machine learning (AI/ML). Further, there occurs dramatic minimization of power consumption and improvement in system performance when traditional algorithms are replaced with deep learning-based AI techniques. Implementation of ML algorithms enables wireless networks to advance in terms of offering high automation levels from distributed AI/ML architectures applicable at network edge and implementing application-based traffic steering across access networks. This has enabled dynamic network slicing for addressing different scenarios with varying quality of service requirements and has provided ubiquitous connectivity across various 6G communication platforms. Keeping a view of the aforementioned, in this chapter, the authors present a survey of various ML techniques that are applicable to 6G wireless networks. They also list open problems of research that require timely solutions.

Chapter 4

Sridevi, Karnatak University, Dharwad, India
Apoorva Shripad Patil, Karnatak University, Dharwad, India

The internet of things describes the connection of distinctive embedded computing devices within the internet. It is the network of connection of physical things that has electronics that have been embedded within their architecture in order to sense and communicate to an external environment. IoT has turned up as a very powerful and promising technology, which brings up significant economic, social, and technical development. Meanwhile, it also brings up various security challenges. At present, nearly nine billion 'things' (physical objects) are connected to the internet. Security is the major concern nowadays as the risks have very high consequences. This chapter presents a detailed view on the internet of things and the advancements of various technologies like cloud, fog, edge computing, IoT architectures, along with various technologies used to prevent and resolve these security and privacy issues of IoT. Finally, future research opportunities and challenges are discussed.

Vaishnavi Shukla, Vellore Institute of Technology, Chennai, India
Atharva Deshmukh, Terna Engineering College, Mumbai, India
Amit Kumar Tyagi, Vellore Institute of Technology, Chennai, India

Cyber security has been an emerging concern of individuals and organizations all over the globe. Although the increasing dependence of the world on the internet proves to be an advancement of technology, it also happens to be a threat to important and private information, monetary fraud leading to huge losses for organizations, and many other issues. In this day and age, cyber security has become a great necessity, and great efforts have been made to enhance it. Artificial intelligence has emerged to be of immense importance over the past decade and is expected to bear great fruit in the coming time. It is no surprise that organizations are depending on AI-driven technologies to protect their data. AI, along with concepts of machine learning and deep learning, is being used to develop new and improved means to help in the cyber security domain. The scope of this chapter covers a few artificial intelligence concepts which have been used in the past to ensure security, a few ideas which have been discussed for future implementation, threats of using AI, etc.

Sridevi, Karnatak University, Dharwad, India
Manojkumar T. Kamble, Karnatak University, Dharwad, India

IoT devices are used to make human life easier and better by saving time and human energy. The IoT devices are controlled by an artificial intelligence system so that the devices can take the necessary decision and perform the work efficiently. The IoT devices are used in homes, offices, factories, schools, traffic signals, water supply management, power sector, security surveillance, hospital, vehicle monitoring, smart city, etc. IoT devices play lifesaving things in human life by continuously monitoring human health. In this chapter, the privacy and security-related IoT device issues were discussed with real-time attacks, and some counter-attacks have been explained. The flow of the chapter is organized as an introduction, architecture, functions, storage management, privacy, security, key elements of IoT, key technologies in IoT, and research opportunities in the IoT domain respectively.

Lipsa Das, Amity University, India
Smita Sharma, Amity University, India
Suman Avdhesh Yadav, Amity University, India
Khushi Dadhich, Amity University, India

The IoT is a current technology that has the ability to interconnect the embedded devices and merge the sensors and wireless networks to communicate and exchange data with one another via the internet. Building and automation, healthcare, monitoring, agriculture, etc. are applications of IoT. Currently, with the rapid increase of IoT applications, we created an environment for device-to-device interconnections, new business models, where these interconnected devices accumulate, process, and share the data with each other. Data integrity or ownership issues, cyber-attacks, single point of failure, etc. are some of the limitations of current IoT solutions. Blockchain technology or a distributed ledger technology has the potential to greatly enhance the security aspect, privacy, and reliability concern of the data. High efficiency, transparency, low cost, and no third-party interventions make this technology a one-stop solution for limitations to integrated IoT devices.

Chapter 8

In recent decades, the industrial applications of the internet of things (IoT) have been attracting massive motivation for research and improvement of industrial operations. The IoT technology integrates various smart objects (or things) to form a network, share data among the connected objects, store data, and process data to support business applications. It is challenging to find a univocal architecture as a reference for different business applications, which can relate to many sensors, intelligence devices, networks, and protocols for operations. Moreover, some of the IoT infrastructural components are a shortage of computational processing power, locally saving ability, and data communication capacity, and these components are very vulnerable to privacy and security attacks. This chapter presents an overview of different IoT-based architectures and security-related issues. Finally, the chapter discusses the challenges of cryptography and blockchain-based solutions after reviewing the threats of IoT-based industry-specific business cases.

Chapter 9

The term "internet of things" is becoming increasingly popular and promising, ushering in a new era of smarter connectivity across billions of gadgets. In the foreseeable future, IoT potential is boundless. The healthcare industry, often known as IoHT, is the most demanding application of IoT. Sensors, RFID, and smart tags are used to start any IoT system, but these applications lack the necessary resources such as power, memory, and speed. The key requirement is secure information transformation because it contains sensitive patient information that might be extremely dangerous if it falls into the hands of an unauthorized person. Encryption approaches that have been used in the past are ineffective. Lightweight cryptography is the most viable solution for protection of data at the physical layer.

Vinodhini Mani, Sathyabama Institute of Science and Technology, India
Kavitha C., Sathyabama Institute of Science and Technology, India
Baby Shamini P., RMK Engineering College, India
S. R. Srividhya, Sathyabama Institute of Science and Technology, India

Game theory and 5G data transferring structures are used to enhance data offloading of computational data. All such solutions not only reduce the load on the cloud by processing edge data and information but also play an important role in privacy and security by ensuring which data communication is locally translated into a network that directly connects the user equipment and then sends the local server to the network core of the organization. With a combination of 5G structure and game theory, we can comfortably handle data transmission, data security, and also system efficiency. The game theory approach with the dynamic computation offloading algorithm (DCOA) allows IIoT (industry internet of things) devices to make decisions spontaneously and reduce computational offloading while transferring data to the local server. The small range of 5G makes it possible to use a small cell network run by a central hub. As a result of its higher frequency, 5G follows a different wireless spectrum structure (N-RAM) and is enhanced by slicing the network.

Sumit Dhariwal, Manipal University Jaipur, India
Avani Sharma, Manipal University Jaipur, India

A new technology approach for improving the efficiency of BB84 protocol for Q-Manet is proposed in this chapter. One of the valuable factors in quantum cryptography is that it ensures vital protection. This chapter explains that the quantum communication for Q-Manet can be successfully sent through both quantum channels and classical channels. The tradition of quantum channel technique for quantum communications on Q-Manet are analyzed, which ensures the way to remove the practical difficulties of secure communication. This chapter provides a new mechanism for improving the communication and thereby improving the efficiency of BB84 protocol quantum communication for Q-Manet.

Chapter 12
Scheduling Optimization Based on Energy Prediction Using ARIMA Model
Pooja Chaturvedi, Institute of Technology, Nirma University,
Ahmedabad, India
Ajai Kumar Daniel, Madan Mohan Malaviya University of Technology,
Gorakhpur, India

Wireless sensor networks (WSNs) have attracted great attention because of their applicability in a variety of applications in day-to-day life such as structural monitoring, healthcare, surveillance, etc. Energy conservation is a challenging issue in the context of WSN as these networks are usually deployed in hazardous and remote applications where human intervention is not possible; hence, recharging or replacing the battery of sensor nodes is not feasible often. Apart from energy conservation, target coverage is also a major challenge. Scheduling the nodes to exist in active and sleep modes is an efficient mechanism to address the energy efficiency and coverage problem. The chapter proposes an ARIMA model-based energy consumption prediction approach such that the set cover scheduling may be optimized. The chapter compares the efficiency of several ARIMA-based models, and the results show that the ARIMA (0,1,2) model provides best results for the considered scenario in terms of energy consumption.

Preface

The Internet of Things (IoT) enables communication environments to support a wide range of applications when equipped with 5G connectivity. As a concern mode of communication to security and privacy, it is susceptible to a variety of distinct kinds of potential assaults, for example, replay, impersonation, password reckoning, physical device stealing, session key computation, privileged-insider, malware, man-in-the-middle, malicious routing, etc. Protecting the infrastructure of a 5G-enabled Internet of Things communication environment against assaults of this kind is thus of the utmost importance. The factors that influence these challenges are coverage, dependability, range, reliability, scalability, security, speed, etc. Because of this, the researchers working in this field are required to come up with a variety of different security protocols that fall under various categories, such as key management, user authentication/device authentication, access control/user access control, and intrusion detection. Therefore, the proposed book effectively helps academicians, researchers, computer professionals, industry people, and valued users.

The influence of next-generation wireless networks, technology and the telecom sector is remarkable on modern society. Sensor networks have become such a critical component in contemporary science and technology; many tasks would not even be possible without them. IoT is a truly interdisciplinary subject that draws from synergistic developments involving many disciplines and is used in the intelligent environment, telecommunication, computer network vision, and many other fields.

In addition to this, a variety of security needs and probable threats in this communication environment are described. The subsequent step is the preparation of the various kinds and categories of security procedures. Additionally structured are the many kinds of analyses of the current security protocols in the setting of 5G-enabled Internet of Things devices. It has gained momentum and popularity as it has become a key research topic in wireless networks. This book has put a thrust on this vital area.

This book focuses on information security practices for the internet of things, 5G, and next-generation wireless networks and its analysis for final-year undergraduate or first-year postgraduate students with a background in engineering, computer intelligence, remote sensing, radiologic sciences or physics. Designed for readers who will become "end users" of wireless communication in various domains, it emphasizes the conceptual framework and the effective use of 5G communication tools. It uses mathematics as a tool, minimizing the advanced mathematical development of other textbooks.

Security and privacy for the internet of things, 5G, and next-generation wireless networks, this book is ideally designed for communication engineers, computer engineers, professionals, academicians, researchers, and students seeking coverage on problem-oriented processing techniques and sensor technologies. The book is an essential reference source that discusses information security practices for the internet of things, 5G, and next-generation wireless networks.

This book is intended to give the recent trends on IoT enable 5G communication for information security and blockchain applications and to understand and study different application areas. This book mainly focuses on stepwise discussion, exhaustive literature review, detailed analysis and discussion, rigorous experimentation results and an application-oriented approach.

Biswa Mohan Sahoo
Manipal University, Jaipur, India

Suman Avdhesh Yadav
Amity University, India

Chapter 1
IoT–Enabled 5G Networks for Secure Communication

Sridevi
Karnatak University, Dharwad, India

Tukkappa K. Gundoor
ⓘD https://orcid.org/0000-0002-6102-0094
Karnatak University, Dharwad, India

ABSTRACT

Internet of things (IoT) meets 5G communications, which aims to use a variety of promising network technologies to support a significant number of connected devices. For cognitive computing, massive machine type of communication (mMTC), cloud computing, artificial intelligence (AI), and so on, huge challenges for security, privacy, and trust are predicted. Technologies for 5G wireless communication encourage its use of mobile networks not just to connect with people and machines but also to connect and manage other equipment that supports virtual reality, such as self-driving cars, IoT drones, surveillance, and security. It is also critical to safeguard the technology of the 5G networks for IoT communication from threats. The different models for 5G IoT communication environment, vulnerabilities, attacks, and its several security protocols are described. The current security mechanisms in 5G networks IoT nature were analyzed and compared. The security challenges and future orientations of 5G-based systems are discussed.

INTRODUCTION

IoT is a network of items that are physically connected such as computers, machinery,

DOI: 10.4018/978-1-6684-3921-0.ch001

and automobiles that are given unique identifiers and may communicate, compute, and coordinate at any time and in any location. It can offer access to the state of things for real-time business, computing devices, equipment, and people, as well as data transport through network. It doesn't necessitate human-to-compute or human-to-human communication, rather than encouraging and expanding machine-to-machine (M2M) connectivity. IoT has had rapid applications in recent years, and it will continue to drive technology for the foreseeable future. The Internet of Things holds immense promise for enabling number of new services and applications, in addition determining the several industries' future. Even as number of IoT devices grows, the current 4G wireless communication will be unable to keep up, efficiency and latency requirements of the internet of things. Cellular 5G wireless was created to suit the needs of future IoT devices, and it has the ability to link some unprecedented various devices. With increased carrying frequency, 5G will constitute a new paradigm, large, unprecedented congestion at the base station and on the devices, and previously unheard-of amount of horns. By 2023, 5G will be ten times faster than 4G Long Term Evolutionary, as shown in a recent Ericsson Mobility Report., with a population of 700 to 3.2 billion people. 5G could revolutionize the Internet of Things.

5G mobile communication technologies, uses the mobile network link and operate machines and other equipment as well as humans (i.e., smart devices). It boosts performance and efficiency, allowing for better user experiences. By delivering 10 gigabits per second throughput, extremely low latency, and high capacity, 5G gives users with consistent services. It is the construction of a wireless network based on "802.11ac IEEE wireless network standard," with the goal of increasing data transmission speeds by three times over the previous "4G-IEEE 802.11n" standard. The enhanced Mobile Broadband (eMBB) technology is bringing 5G Internet based material (IoT) to life as a smart house and smart building with the debut of 5G. The 5G Non-Standalone (NSA) network is currently available globally, has enabled Mobile Broadband (eMBB) technology. With the introduction of high-tech (mMTC) technology and extremely-reliable Low Latency Communications (URLLC) era, the 5G Standalone (SA) network changed into first brought inside the United States of America and China, and has grown to make IoT extra widely available. New traits emerge as IoT gadgets integrate 5G networks, such as the EMBB smart home, however with better safety issues. If IoT gadgets have inadequate protection linked to the 5G network, they may be less probable to pay attention and keep away from fees because of device risks. risk studies and the layout of the IoT tool to boom this chance are crucial for 5G-based IoT age. (R. Khan. et al. 2020; A. Ahad. et al. 2019; A. Braeken. et al. 2019; Ericsson. 2019; Qualcomm. et al. 2019).

This chapter, begin by evaluating 5G and IoT, as well as the characteristics of mobile communication in 5G, the benefits and drawbacks of IOT, and the convergence of 5G and IOT. 5G in IoT applications, system models, 5G in IoT communication security threats, Threat assessment for 5G in IoT, Protection from IG 5-enabled IoT attacks, security protocols of 5G IoT network protocols Challenges and future research guides draw attention to new concerns in IoT-enabled 5G challenges. It also compares 5G IoT-enabled security to privacy concerns, demonstrates and analyzes modern security and privacy across edge paradigms, and identifies potential interactions between security measures in the 5G in IoT nature.

BACKGROUND

2G/3G/4G, Wi-Fi, Bluetooth, and other wireless technologies are used at various IoT applications, where billions of physical devices are linked via wireless network technology. 2G networks (presently covering 90%) are intended for use with voice, 3G networks (currently covering 65% of the world's) are designed for voice plus data, and 4G networks (available as of 2012) are highly developed. Speed internet. Although IoT is gaining popularity, 3G and 4G networks aren't yet fully developed of internet of things applications. (Akpakwu, et al. 2017). The capability of mobile networks that provide internet connectivity to IoT devices has been greatly improved due to 4G. Since 2012, 4G connectivity through Long Term Evolution (LTE) has evolved into the quick and most consist form 4G cellular technology compared to other technology such as Bluetooth Low Energy (BLE) (BLE, Smart Bluetooth Low Energy), WiMaxb (L. A. Taylor. et al. 2011), ZigBee (Tech. Rep., 2016), and SigFox (RPMA. 2016). Vangelista et al. (2018) must fulfil the requirements of smart environment, industry 4.0 and other recent technologies, 5G networks and standards are planned to address the complex connectivity, device integration and intelligence that 4G networks face.

MOBILE COMMUNICATION SYSTEM EVOLUTION

Evolution of the latest generation of overall technology (3G, 4G or 5G), and analyse the differences in mobile communication across time. The detailed explanation of the evolution of mobile communication systems are (Costanzo and Masotti, 2017):

1. **Previous generations (1G, 2G and 3G):** In the late 1970's, 1G (First Generation) was introduced. It did not have strong security features, had phone drop problems, and had a data transfer speed of about 2.4 Kbps. Data services such as short message service (SMS), Multimedia messaging service (MMS) are supported by 2G (Second Generation). Data transfer speeds via General

Packet Radio Service (GPRS) ranged from 50 megabytes to one second per second. Web browsing, email, movie downloads, photo sharing, and other smartphone apps were all available in 3G (third generation). It has a fixed 2 Mbps data transmission rate and a mobile data transfer speed of 384 kbps. High speed packet access (HSPA +) has a speed of about 21.6 Mbps.

2.   **Fourth Generation 4G:** It provides customers with more quality, high power and fastest services. Provides better security, cheaper information and voice services and the Internet via IP. It can be used for advanced mobile Internet access, games, high quality mobile TV, IP telephony, video conferencing, 3D television and other designated computer sites. It uses Multiple Input Multiple Output (MIMO) technology and Orthogonal Frequency Division Multiplexing (OFDM) technology.

3.   **Fifth Generation 5G:** Designed to improve 4G boundaries and performance. It has high connectivity and a fast data transfer rate with minimal latency. It facilitates communication between device and device while consuming less power and provides better telephony. 5G data transfer faster than 4G, which can reach a maximum of 35.46 Gbps. Technologies running on the theme of Multi Input and Multiple Output (MIMO) and Millimeter Wave Communications. Technologies and technologies such as small cell, large MIMO, millimeter wave, and light reliability (Li-Fi) are used to give 10Gbps with very little latency. It allows connectivity across nearly 100 billion gadgets. (Accessed: Dec. 2019. Online, Ahmad et al., 2018, Gandotra. et al., 2017).

Figure 1 depicts the progression of mobile communication systems over the previous few decades. It also offers information on the functionality characteristics that are supported by different generations.

The four megatrends mentioned above are included in 5G. (E.g., increase in number of devices, increase in traffic, a greater reliance on the cloud, as well as a variety of 5G convergence services). Some interested companies have made proposals to determine the key performance metrics that will be used in the 5G network. The international telecommunication Union-Radio Communications sector(ITU-R)(fifth generation of mobiles) based its recommendations on some of the key criteria set out in Table 1(Chen et al., 2020)(Hao, 2021)(Fuentes et al., 2020).

5G CHARACTERISTICS IN MOBILE COMMUNICATION

The 5G communication delivers faster and more dependable connectivity in smart phones and some other devices such as smart vehicles etc.

Figure 1. Mobile communication system evolution/generations (adapted from (Tian et al., 2019).

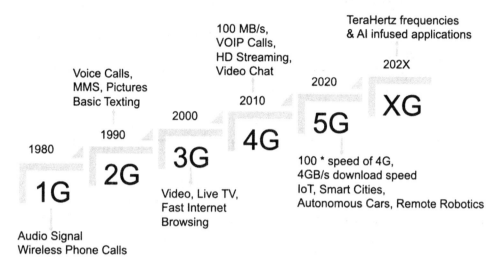

- **Top speed**: Data travels a hundred times quicker in the 5G network than before. At full expansion, data transfer speeds of up to 10 Gbps are possible. As semiconductor technology advances, a compact chip can hold more and more power.
- **Networks**: 5G is capable of supporting a large number of wireless connections. This is critical for the Internet of Things, as billions of stationary and mobile things will be able to communicate with one another. The capacity and throughput of data have improved.
- **Delay-free communication**: Response times of less than one millisecond is considered almost real-time. This kind of delay is required for a variety

Table 1. 5G's key performance indicators.

Key performance indicator	Target Value
Peak data rate on the downlink	20 Gbps
Peak data rate on the uplink	10 Gbps
Anywhere in the Cell, Average Data Rates	100 Mbps
Device per km2	1 million
Area capacity	10 Mbps/m2
Over-the-air latency in User Plane	10 ms roundtrip
Over-the-air latency in Control Plane	10 ms

of advancements, including autonomous driving, telemedicine, and other manufacturing and service activities. Virtual and augmented realities are also aided by the technology.

- **Low energy consumption**: For data transfer, 5G uses less energy per bit than prior standards. This, according to Huawei, can save up to 90% on electricity. This is appealing not only to smartphone users, but also to Internet of Things and other narrowcasting applications.

BENEFITS AND DRAWBACKS OF INTERNET OF THINGS

Any technology available today has not yet attained its full potential. There's always a void to fill. As a consequence, we can claim that the Internet of Things is a significant technology across the globe that can assist other technologies in reaching their full accuracy and potential. Some of the primary Benefits and drawbacks of the Internet of Things.

1. **Benefits of Internet of things:** In area of commerce, the IoT offers many benefits in everyday life. Some of its benefits are as follows:
   a. It can be used to control smart house and cities with cellular phones. It boosts privacy and protects personal information.
   b. Task automation saves us a lot of time.
   c. Even if significant distance from a actual location, information is readily available and is regularly updated in real time.
   d. Electrical devices such as cell phones are connected directly to a controlling computer and communicate with it, resulting in efficient power consumption. Unnecessary electrical equipment should not be used.
   e. IoT apps can provide personalized support by alerting general plans.
   f. It reduces human labour by allowing IoT devices to link and interact with each other and without the need for human involvement, execute a number of activities.
   g. Patient care can be done more effectively in real-time without the need to visit a 'doctor'. It allows them to make decisions while providing evidence-based care.
   h. Asset monitoring, travel or transport tracking, asset tracking, distribution, and surveillance are all examples of asset tracking personal order tracking and customer management are all reduced with a proper tracking system.
2. **Drawback of Internet of Things:** As the Internet of Things provides a number of advantages, it also has a number of challenges:

a. Hackers can be obtaining access to the network and steal sensitive information. Because it connects so many things to the internet, user data is likely to be used against it.

b. Users are totally dependent on the internet and cannot function properly without it.

c. Due to IoT complexity, systems can fail in many ways.

d. User may lose control of lives and become completely dependent on technology.

e. Excessive use People are influenced by the internet and technology illiterate Instead of doing things by hand, they equipped with smart equipment, making them lazy.

f. Managing, building, Planning, and enabling a broad technology to IoT framework is quite complex.

5G AND IOT CONVERGENCE

The confluence of 5G technologies in the Internet of things is the logical to other step for two cutting-edge model designed to make user lives easier more convenient and more productively. It discussed how they combined and comprehend each of the two methods. Simply defined the next generation telecommunications network is known as 5G. In comparison to 4G, the current standard, which delivers upload rates of 7 to 17 Mbps and download speeds of 12 to 36 Mbps, 5G transmission speeds could reach 20 Gbps. The latency of 4G transmission will be close to 10% and the number of gadgets that can be connected is expected to skyrocket, necessitating IoT convergence. When both techniques are combined, 5G would have a significant impact on all IoT components: sensors also have more frequency band to report actions, networks will deliver more data efficiently, real world data will be a reality for cloud and AI, and applications will get more features and cover a wider range of options due to the increased bandwidth provided by 5G.

By 2021, 20 billion IoT devices are expected to be linked to the worldwide technology, which will generate huge amounts of information. According to the researchers, the 5G in IoT model is a combination of 5G networks, Artificial Intelligence (AI), and IoT to create the IoT (I-IoT) environment. A combination of three such technologies provides users to have effective accessing to data created by IoT devices, which can be used to understand data. Compared to existing cellular network communication systems, the development of 5G become more important for IoT expansion. 5G allows for high speeds, higher coverage and larger broadband. IoT devices (for example, IoT sensors) are linked to a cloud server that collects sensory data (Lee, 2015; Wazid et al., 2018; Vasilakos. et al., 2018; Rodrigues et al., 2019).

5G IN IOT APPLICATIONS

5G IoT will improve the quality of life of daily users. Facilities will continue to improve with 5G IoT, allowing crucial upgrades to be sent to whole networks without freezing functionality, interrupting operations, or overloading servers. The following industries will continue to benefit from 5G IoT advancements:

1. **Automotive:** One of the basic usage cases for 5G is the connected car, which includes both direct linked (between vehicles and pedestrians, infrastructure) as well as network-facilitated communications for autonomous driving. The applications are real-time local updates focusing on intent sharing, path planning, coordinated driving and vehicle comfort and safety. Edge computing, a promising cloud computing derivative, allows IoT devices to perform processing, decision-making, and action-taking while only sending relevant data to the cloud.

2. **Remote surgery:** Reduced latency is one of the most important advantages of 5G. Because there is a very small the time it takes for a device to ping the network and receiving a reaction, as compared to 4G LTE, when this was a problem. A specialist may operate remotely, surgery without having to be present in the same operating room thanks to the characteristics of the 5G network. Doctors at King's College London, used a dummy patient using "virtual reality headgear" and a "special glove" to demonstrate a surgery procedure. As a result of this, they were able to do remote surgery using a robotic arm. (Y. Park et al. 2019).

3. **The Industrial Internet of Things (IIoT):** It's a physical item, systems, platforms and uses with embedded technology that can interact and share intelligence with each other, the outside world, and humans. The IIoT's adoption is aided by the increased Sensor accessibility and costs, CPUs, and other network that have made real-time data gathering and access easier. Private 5G networks will see a lot of use from industrial enterprises. (TrendMicro accessed on 2021).

4. **Self-driving car:** It's also known as a self-driving car (AV) is a vehicle that is smart perceives its platform and navigate without requiring much human involvement. Sensors of various types are fitted in self-driving automobiles to gather information about their surroundings. Other important features included in these vehicles include inertial measurement units, radar, odometry and GPS system. The installed control system can evaluate the required travel courses as well as the detected data to identify any obstacles that may arise. AVs can also communicate with one another in order to make informed decisions and

choose the best approach to a destination. (R. Hussain et al. 2019; M. A. Imran et al 2019; Jo and Sunwoo et al. 2014).

5. **Healthcare**: 5G will meet the demand for real-time networks, which will have a significant impact on the healthcare industry. One application is the remote monitoring of slightly elevated operations videos that are streamed live. The concepts of real-world and higher frequency Telemedicine will become a reality, and Smart applications will become more advanced, allowing for more in-depth health data about individuals to be collected on the go. The epidemic of 2020 showed us the value of seeing our doctor in methods other than in person, and many entrepreneurs developed telemedicine apps that time. 5G will accelerate the adoption of some apps, making doctor appointments more efficiently and with less time waiting. (Computer weekly accessed on 2021).

6. **Virtual reality:** It's a type of man-made environment made possible by the use of software and other technologies. It is given to the user suspends belief that the user will be interested in learning more about it. forget its reality and it is accepted as real world. It is mostly experienced through the senses of sound and sight. How 5G could expand the user's experience was illustrated at a recent mobile world congress. It allowed users to interact with live-streaming virtual worlds in real time. (MWC. Limitless Intelligent Connectivity. Accessed: Oct. 2019. Online, M. Shafi et al 2017).

7. **Flying IoT Drones:** The Internet of Devices (IoD) is a layered network control architecture that allows human and autonomous flying devices to communicate with one other (i.e., drones). It's employed to manage the air space and provide services for navigating into various locations. Traffic monitoring, search and rescue, and package delivery are just a few of the services provided by flying IoT. Those drones use an Internet connection to interact with the access point (server)(S. L. Waslander. et al. 2016; J.J.P. C. Rodrigues et al 2019).

5G IN IOT COMMUNICATION POTENTIAL ATTACKS AND SECURITY

Various security and privacy issues are related with 5G networks. Internet of things connectivity, potential risks, and the need for security (H. Gharsellaoui. et al. 2017; H. Almagwashi. et al.2018; W. Shi. et al. 2019; H. Zhao. et al. 2017, M. Guizani. et al. 2020).

1. Issues with privacy and security in 5G networks IoT interactions Despite of the fact IoT is rapidly developing, researchers have devised innovative approaches

to addressing IoT connectivity difficulties, however it continues to face security and privacy issues.

a.   **Openness of the network:** Connecting IoT devices to a user network that is linked to other systems is critical. It's also likely that some IoT devices have security weaknesses that could harm the user network. This is because it serves as an entry point to get access to the system via hackers

b.   **Sensitive data privacy:** The Internet of Things includes a huge number of gadgets, each with its own software and hardware. Replay, MITM, cheating, password guessing and other attacks can be used against some of them. Unauthorized access and manipulation may result in the disclosure of sensitive data. Some devices send personal data such as username, address, birth date, mobile number, credit card details, and patient records are all required. Protecting IoT communication from potential attacks is always required.

c.   **Inadequate security measures:** IoT devices, such as a PC or smartphone, are connected to the system. In such a situation, the lack of a security raises the risk of personal information leaking. Data transmitted and collected through IoT devices are available to the general public (i.e., health information collected via smart health devices).

2.   Requirement for security in IoT ecosystem with 5G connectivity It describes the several security needs in a 5G in IoT ecosystem, as well as the generic security requirement.

a.   **Authentication procedure:** In this approach is used to verify communication device identities (i.e., IoT device). Before commencing secure communication, mutual identity verification is essential, and it should be completed ahead of time. In a 5G-enabled IoT ecosystem, smart IoT devices, different types of servers (servers, fog, edge cloud,), various types of users, service gateway nodes and providers may all require authentication.

b.   **Integrity property:** It guarantees that the information is real and correct. The received message's content should not allow for unlawful inject, unauthorized change. Data should be kept confidential.

c.   **Confidentiality property:** This attribute protects data from being accessed from unauthorized users. It is also known as privacy in another sense, as it shields the exchanged messages against attacks on information.

d.   **Non-repudiation property:** It guarantees that no entity will discard the validity of the message. This is an important service in information security because it proves the source of the text and the integrated of the data in that message, making it impossible for illegal organizations to dispute the source of communication and authenticity. Digital signature methods are important for denial. The report is categorized into the following:

    i. **Non-repudiation of the source:** It serves as proof of the sender's authenticity. This establishes that the message was sent from a trustworthy source.

    ii. **Non-repudiation of the destination:** It guarantees the receiver's legitimacy. This verifies that the message was delivered to the correct person.

e. **Procedure for Authorization:** The authorization method is used to restrict access to the device or user system resources network (i.e. access restrictions) (i.e. data applications, files and services). Authentication techniques are designed to validate the identity of devices or users in general. The rules for gaining access to all kind's resources are created and implemented by an authority (such as an administrator).

f. **Property of freshness:** This feature assures that the data is "fresh." Invalid entity will not replay messages already sent.

g. **Property of Availability:** It ensures that only genuine businesses have access to the information. If an adversary is unable to attack the data's confidentiality and integrity, the user may resort to other destructive measures, such as launching a DDoS (Distributed Denial-of-Service) assault on a web server that brings all related websites down.

h. **The property of forward secrecy:** If an entity leaves the communication network (for example, a smart IoT device or smart home it must lose access to future conversations.

i. **The property of backward secrecy:** If an entity (such as a smart house user or a smart IoT device) is connected to a communications network for the first time, it may not able to view old sent messages.

3. Potential 5G attacks in IoT The following potential assaults on a 5G-enabled IoT communication could be carried out by active or passive adversary. (Das and Zeadally, 2019):

a. **Attack based on impersonation:** In this malicious operation, an attacker accurately establishes a person's identity valid communicating party and then generates and transmits a communication sent on behalf of the authentic communication person to the receiver.

b. **Distributed Denial-of-Service Attack (DDoS):** An adversary engages in this harmful behaviour to obstruct lawful parties from accessing system or network resources (For example, some data resources or some IoT devices). Another type of DoS assault is the Distributed DoS (DDoS) attack, which is carried out by many attacker systems at the same time. Flooding attacks against UDP, HTTP, and TCP SYN are a few examples.

The flooded packets in these attacks quickly drain the targeted system's resources (for example, bandwidth) (i.e., web servers). (S. U. Jan et al. 2019; H. Chen et al. 2019; Kamaldeep. et al. 2017) DoS attacks can also be carried out through 5G-enabled IoT communications through the use of various forms of attacks on routing such as, greyhole, sinkhole, blackhole, wormhole and misdirection attacks. In the midst of some attacks, data transfers (packets) might be dropped, delayed, altered before they arrive at their destination receiver. This increases end-to-end delay, decreases throughput has an effect on other network performance characteristics. (M. Wazid et al. 2017; M. Wazid et al 2016) Formalized paraphrase.

c. **Eavesdropping:** Also known as a sniffing or snooping attacks. happens when an attacker listens in on conversations between communication parties. Because it allows the attacker to launch more attacks, one of the most likely attain 5G in IoT connectivity.

d. **Traffic analysis:** It's a passive attack wherein the attacker intercepts and to explain what's going on, the communications are analyzed.

e. **Attack on Replay:** When a hacker hijacks messages, it's called eavesdropping and then erroneously delays or retransmits them in order to confuse the recipient entity.

f. **Man-in-the-middle attack (MITM):** In this harmful conduct, the attacker seizes the transferred communications before attempting to change or delete them before sending them to the intended recipient.

g. **Database attack:** Database-related attacks on databases handled by separate servers, such as a fog server or a cloud server, are also possible in a 5G-enabled IoT communication. Three examples include a Cross Site Scripting (XSS) attack, a SQL attack, and a Cross Site Request Forgery (CSRF) attack. The current opponent conducts these attacks in order to do financial harm. User may make an illicit money transferring from a genuine user account or alter a legitimate user's password.

h. **Malware attack:** An adversary may run or execute a malware on a remote server to do illegal actions such as stealing, destroying, updating, and encrypting sensitive information. Malware can take several forms, including as viruses, worms, key loggers, ransomware, spyware, Trojan horses and. They're also utilized to keep track of things. user activity without their knowledge. In the IoT setting, botnet can be used to spread hostile malware (affiliate and collaborative attack systems). Botnets are Echobot, Reaper, Mirai, and Neckers are active in these days. Smart device can be hacked remotely by malware (control)(T. S. Messerges et al.2002; R. Khan et al.2020; J. Granjal, et al.2015; M. A. Khan, et al. 2018; A. K.Das, et al. 2018)

i.   **Intrusion by insiders:** In this malicious behaviour, a trustworthy authority's high-ranking inside client uses the data saved to carry out further essential assaults, such as stealing compute key exchange passwords.

j.   **Capture of deploying devices physically:** IoT device monitoring physically not possible to be available 24 hours a day. When an opponent has the opportunity to physically capture these devices, the hacker will attempt to retrieve sensitive information from their memories (such as identities, secret keys, and so on). In a IoT 5G-enabled context, an adversary may use this information to perform a range of undesired activities (Such as attire, session key computation, password guessing and man-in-the-middle (MITM) attacks) (W. Zhou et al.2019).

SECURITY PROTOCOL CATEGORIES IN A 5G IOT COMMUNICATION ENVIRONMENT

5G in IoT communications nature is vulnerable to a variety of assaults. Researchers in this sector presented several security protocols, which can be classified as key managing protocols, Protocols for verification identification, control security systems, and intrusion prevention. (Messerges et al., 2002; Wazid, et al., 2016, 2017, 2019)

1.   **Intrusion Detection**: Intrusion prevention methods collect and evaluate harmful activity within a system or network. The term Intrusion detection system (IDS) refers to a system that performs the function of intrusion detection. An intrusion detection system (IDS) protects several devices, some smart (IoT) devices, against potential threats. In a 5G in IoT environment, the intrusion detection technique employed examines and validates different forms of traffic (which could be malicious or normal), and then forecasts the indication of incursions. If an intrusion is discovered, the related programme performs the appropriate actions, such as blocking the malicious source's IP address or notifying the administrator of the breach. In addition, the opponent is likely to physically steal some IoT devices. The opponent may use a subsequent Power Analysis Attack to attempt (Challa et al., 2018) to obtain confidential data (i.e. evidence, private keys) from the stolen device. By storing crucial information in these nodes the adversary tries to establish his malicious nodes (s). Additionally, IoT botnets can target this type of communication to infect IoT devices operating systems or memory with malware. In the presence of Malware assaults IoT gadgets could malfunction or stop working completely. It is critical to safeguard 5G in the IoT context from intrusions. Strategies for detecting intrusions are essential for such environments. (M. Wazid, et al.2018;

Figure 2. Phases of key management protocols (adopted from(Wazid et al., 2020)).

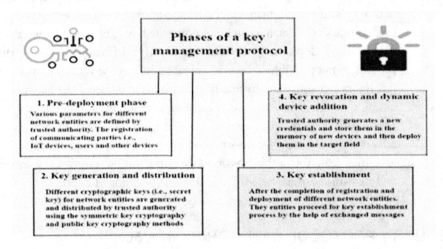

Magellan solutions accessed: Oct. 2019; Z. Liu, et al. 2019; J. Su, et al.2018; V. Clincy, et al.2019; Y.-W. Kao, et al. 2013; J. Li, et al.2014; P. K. Tysowski, et al.; J. Yu,et al.2015; L. Eschenauer, et al. 2002).

2. **Key management:** In an IoT environment, a key management protocol generates, distributes, establishes, and manages cryptographic keys among communication parties. According to the demand, the defined procedure includes phases are key exchange, key production, key usage, and key revocation. this protocol employs a cryptographic technique to keep track of key servers (the trusted authority), different users (stationary or Phone) and the devices they utilize (for example, smart device). Very Strong key management practices must be followed to guarantee secure communication. (H. Chan, et al.2003; W. Du, et al.2004; W. Du, et al.2003; C. Blundo, et al.1993). A key management protocol will typically include words such as pre-deployment phase, key creation and Phases of dissemination, fulfillment process, and key revocation and dynamical device addition. (Y. Cheng, et al. 2005; D. Liu, et al.2005; Q. Dongand D. Liu.2007; S. Zhu, et al. 2006; A. Perrig, et al.2002; M. Wazid, et al.2017,2016,2018; D. He, et al.2018). Figure 2 describes the various phases of key management procedures.

3. **Access/user control:** The access control methods limit a user's or device's access to the resources of a system or network. The procedure that follows grants different users or devices access to and rights over the various resources that are available. Additional devices, such as smart (IoT) devices, must be

linked to the network to increase the overall lifespan of the IoT communication environment. This could occur if a device ceases to function as a result of battery depletion or physical theft. An adversary may also try to install a malicious device in the target region, and distinguishing between malicious and authentic devices is crucial. Safe access control solutions should be created to prevent hostile organization from entering the 5G-enabled IoT ecosystem. (L. Wu, et al.2019; D. Wang, et al. 2018; M. Wazid, et al.2019)

4.   **Authentication of users and devices:** A strategy for identifying and validating the identities of communication entities, such as a user or a device, is known as person/device authentication. Mutual authentication is the mechanism through which communication entities such as users and smart IoT devices verify one other's identities. The distributing entities generate a session key to secure their connection after completing mutual authentication. The same procedure is used for device authentication. The user authentication method is described in depth to make things easy for you. The registration phase, device-server authentication phase, and device-device authentication phase are the phases of a user identification approach for 5G in IoT network. The intricacies of these phases are depicted in Figure 3. Two factor and three factor user identification approaches, which controls privacy depending on the factors provided, are fairly common in practice. The user's credentials are one of the three factors used in identification methods (i.e., information about the login and password), the device used by the user (i.e., smart card, smartphones), and information about the user biometrics (i.e., her or fingerprints or his iris scan). (Wazid et al., 2018, 2016; Kumar et al., 2019) respectively.

THREATS FOR 5G IN IOT

Threat to the 5G IoT Middleware architecture, as well as security countermeasures and IoT threats combine, are new potentially bigger security issues to consider as part of the 5G IoT Middleware design.

Threats for 5G IoT Middleware architecture

•   **Message modification:** The intruder updates the packet header address in the message editing attack to send it to the new destination or modify the data on the target machine. Among email-based attacks, message manipulation attacks are widely used. The attacker may take advantage of security vulnerabilities

Figure 3. Protocols for user authentication and device authentication in phases (adapted from . Pham et al.2019)

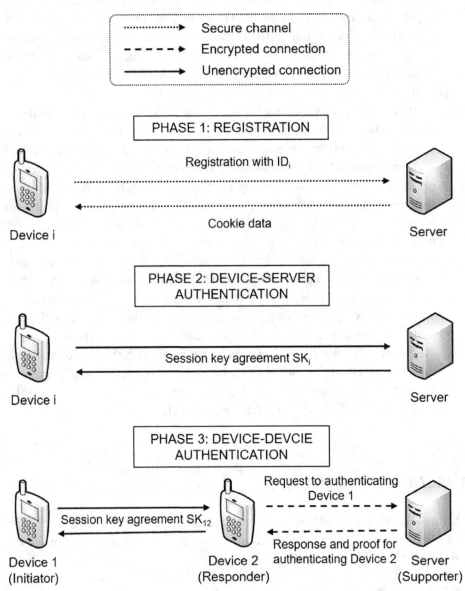

in the email protocol to insert malicious content into the email message. The attacker may insert malicious code into the message body or header data.

- **Man in the Middle:** When a criminal insert himself into a conversation between a user and an application, disguised or pretending to be someone in the party the Man in the middle (MITM) attacks, gives an impression that the conversation is normal.

- **Authentication attacks:** Using the automated trial and error technique, the attacker can guess the username, credit card number, password and cryptographic key.

- **DOS attacks:** Enables the attacker to use an automatic error and trial and technique to guess the password, user's username, credit card number or cryptographic key. Denial-of-service (DoS) An attempt is made to shut down a system or network that isn't accessible to the intended users. DoS attacks attack targets by filling in traffic or information that could cause the target to crash. DoS attack, in any case, end users (employees, members or clients) lose their expected service or resource.

- **Physical attacks:** Physical attacks (sometimes called kinetic attacks) are deliberate hostile actions to damage, alter, expose, stop, gain unauthorized access or steal to physical assets, such as infrastructure or hardware or connections.

- **Replay messages:** Replay attack by listening to a hacker on a secure network connection, intercepting it, and then fraudulently delaying or replaying to deceive the receiver that the hacker wants.

- **Masquerade Attack:** is used to gain unauthorized access to personal computer data through fake identity, legal access identification such as network identity. If the authorization process is not completely secure, it is particularly vulnerable to masquerade attack.

- **Passive attacks**

The goal of a passive assault is to gather information about the system being targeted; no direct action is taken against the target.

- **Traffic Analysis:** The two most common forms of passive assaults are traffic analysis and message content leaking. The eavesdropper analyses the traffic, detecting its location, identifying communicating hosts, and noting the frequency and length of messages sent during a traffic analysis assault.

- **Eavesdropping:** Eavesdropping is the act of surreptitiously or stealthily listening in on another person's private discussion or communications without their agreement in order to obtain information.

PROTECTION AGAINST ATTACKS FOR 5G IN IOT

To secure the entire 5G in IoT middleware architecture from these attacks, security defenses must be implemented in the middleware architecture. The detailed safety guidelines that must be followed to protect the entire structure are:

- **Confidentiality:** Cryptographic techniques can be used to meet this demand. Various existing asymmetric and symmetric cryptography technologies can be utilized to guarantee confidentiality. Because IoT devices work in a limited resource system. The choice of a cryptographic algorithm is heavily influenced by device capabilities. Confidentiality should be implemented throughout the middleware design to protect the transferred data. It can also ensure that an entity's data is safe from unauthorized access.
- **Authorizations:** It's a way of creating access policies that expressly grant particular permissions to subjects based on previously authenticated credentials, reusable, Fine-grained, dynamic, simple-to-use policy development and modification frameworks must be available in the IoT ecosystem. vital to externalize IoT service policy formulation and enforcement mechanisms. In an IoT middleware, authorization is required since apps and devices have various credentials for to accessing different services and resources.
- **Authentications:** Two entities must create a genuine relationship in order to reliably change keys and data. Because IoT data is used in many actuation processes and decision-making, mutual authentication is required in the IoT context. Both parties must have confidence that the service is used by genuine people and is provided by a genuine source. Strong authentication measures must also be developed to prevent impersonation. The registration of user IDs is required for any authentication approach to be enforced. Furthermore, the resource constraints of IoT gadgets place severe limitations on any authentication method.
- **Access control:** This is a security feature that provides only authorized users to access resources. Implementation is usually based on access control options. As the Internet of Things grows in popularity, the issue of privacy has become a big concern. It is vital that only authorized parties have access to user data.
- **Communication channel protection:** This requirement's objective is to ensure communication linkages between devices/apps and middleware safe. The purpose is to safeguard data shared by organizations from assaults during transmission by creating security procedures that must enable network communication protection regardless of safety criteria in place.

- **Integrity:** Through data validation and verification, this criterion assures that an interchanged message has not been tampered with during transmission by an unauthorized person. IoT companies share important data with other enterprises, enforcing a strict demand that data be detected, stored, and transferred without being tampered with, either intentionally or inadvertently. For the development of strong IoT applications, device data integrity protection is essential. This is ensured via Message authentication codes (MACs) that use one-way hash techniques. Again, the device's capabilities dictate which MAC method is used. Integrity can also be used to protect data held by entities.

- **Availability:** In order to give continuous information, Internet of Things services must be accessible from anywhere at any time. This requirement can't be met by any one security protocol. However, there are a few practical steps that may be taken to ensure availability.

CHALLENGES AND FUTURE RESEARCH DIRECTION

The IoT communications platform supports a wider range of applications, including 5G enabled smart housing, industrial automation, e-health, smart metering, and sophisticated robotics. Need for real-time data processing and access, which generates a vast volume of data, a technique should be employed to find any special patterns. It is subject to conventional security, privacy, and other concerns. This section of the research looks at some of the domain's current difficulties as well as potential research directions.

1. **Existing protocol's security**: The majority of IoT security measures are vulnerable since they do not provide complete protection against potential attacks. Furthermore, some existing protocols are designed to support only one attack at a time and do not support many attacks at the same time. It is vital to create security systems that can protect against several threats at once. Future scholars will face a challenging task in designing such protocols for this subject. (B.Wu, et al. 2019).

2. **Efficient security protocol design:** 5G-enabled Resources-constrained devices are IoT sensors with limited computational capability, low storage size, and a small battery unit, are part of the IoT ecosystem. These gadgets are unable of completing activities that require more strength in these dimensions, such as communication, computation, and storage. It is preferred to design secured protocols in such a way that they consume little computing power, have minimal

communication costs, and have a small storage footprint without jeopardising system security.

3. **Security protocol scalability:** 5G in The Internet of things ecosystem is a complex web of communication protocols and applications. Each of these apps has its own set of features and issues to deal with. It will be tough to design a privacy protocol for this type of communication nature in such circumstances. Patient electronic health records, for example, must be collected on a cloud server for decision-making and processing in a smart healthcare communication system. A body area network (BAN) consists of a collection of devices that create information and send it to a cloud server and forms multi-device communication platform.

4. **Data storage privacy:** Data privacy is concerned with the proper handling of data across multiple resources such as compliance, notices and regulating responsibilities. The 5G for IoT ecosystem is also used in sensitive-data processes (such as Smart Health). In this security-sensitive environment, health uses smart gadgets are to track health issues inside / outside the patient's body. After that, comprehensive and sensory health information is sent to cloud servers for storing and processing. Typically, different prospective attackers can have an impact on this type of communication environment. This could result in data leaks in transit as well as data stored on servers. It is very critical to protect data privacy when it's in transit and when it's stored. more efficient procedures are desperately needed to protect data privacy.

5. **Device heterogeneity:** 5G connectivity from multiple processors, laptops, and personal computers, and desktop computers to Resources-constrained sensing devices and RFID tags, the Internet of things encompasses a widely used range of devices. Aside from that, these devices are work with a wide range of communication methods. Communication strength, computing capacity, storage space, and system software installed all varies amongst devices (e.g., Operating system(OS)). Security methods must be designed to protect a wide range of devices, as well as the technologies and procedures that go along with them (Wazid, M, et al.2020; Pham, et al. 2019).

6. **Designing security protocols based on block chains:** The 5G-enabled IoT ecosystem could be protected using block chain activities. Block chain steps are decentralised, efficiently, and visible to all communication nature components because they are decentralised, efficient, and visible. In a 5G-enabled IoT context, block chain operations can be used to define security standards. To perform the available action, a blocks containing data required functionality, such as an data message or authentication message, can be created and added to the block chain. Because block chain has been made available to lawful

network organisations, this entity can retrieve data using the blocks of the block chain.

CONCLUSION

5G-enabled as it is vulnerable to numerous forms of attacks, the IoT ecosystem suffers from a variety of security and privacy challenges. It becomes critical to safeguard, against various security protocols exist to prevent the infrastructure of a 5G in IoT ecosystem from these attacks'. The terms user access control and intrusion prevention are interchangeable were coined under several titles (for example key management, Access control/ user authentication, user authentication/device authentication and Various security issues are discussed in this chapter. In communication environment, the requirements and attacks are the two major factors. There are many security protocol categories are described. Finally, some potential problems in the security of 5G in IoT environments are presented to aid academics working in the field.

ACKNOWLEDGMENT

Mr. Tukkappa K Gundoor is thankful to the Department of Science and Technology of Karnataka (DST) for supporting our work through Ph.D. fellowship No. DST/ KSTePS/Ph.D. Fellowship/PHY-02:2020-21.

REFERENCES

Ahad, A., Tahir, M., & Yau, K.-L. A. (2019). 5G-based smart healthcare network: Architecture, taxonomy, challenges and future research directions. *IEEE Access: Practical Innovations, Open Solutions*, *7*, 100747–100762. doi:10.1109/ACCESS.2019.2930628

Ahmad, I., Kumar, T., Liyanage, M., Okwuibe, J., Ylianttila, M., & Gurtov, A. (2018). Overview of 5G security challenges and solutions. *IEEE Communications Standards Magazine*, *2*(1), 36–43.

Akpakwu, G.A. (2017). A Survey on 5G Networks for the Internet of Things: Communication Technologies and Challenges. *IEEE Access*.

Alsamhi, S. H., Ma, O., Ansari, M. S., & Almalki, F. A. (2019). Survey on collaborative smart drones and internet of things for improving smartness of smart cities. *IEEE Access: Practical Innovations, Open Solutions*, 7(September), 128125–128152. https://doi.org/10.1109/ACCESS.2019.2934998

and the Future of Telemedicine and Remote Surgery | Digi International. (n.d.). Retrieved May 11, 2022, from https://www.digi.com/blog/post/5g-and-the-future-of-telemedicine-remote-surgery

Baumgartner, M., Juhar, J., & Papaj, J. (2021). Short Performance Analysis of the LTE and 5G Access Technologies in NS-3. *Proceedings of the 16th Conference on Computer Science and Intelligence Systems, FedCSIS 2021*, 25, 337–340. doi:10.15439/2021F62

BLE. (n.d.). *Smart Bluetooth Low Energy*. http://www.bluetooth.com/Pages/Bluetooth-Smart.aspx

Blundo, C., Santis, A. D., Herzberg, A., Kutten, S., Vaccaro, U., & Yung, M. (1993). Perfectly-secure key distribution for dynamic conferences. In Lecture Notes in Computer Science: Vol. 740. *Advances in Cryptology* (pp. 471–486). Springer.

Braeken, A., Liyanage, M., Kumar, P., & Murphy, J. (2019). Novel 5G authentication protocol to improve the resistance against active attacks and malicious serving networks. *IEEE Access: Practical Innovations, Open Solutions*, 7, 64040–64052. doi:10.1109/ACCESS.2019.2914941

Challa, S., Wazid, M., Das, A. K., & Khan, M. K. (2018). Authentication protocols for implantable medical devices: Taxonomy, analysis and future directions. *IEEE Consum. Electron. Mag*, 7(1), 57–65.

Chan, H., Perrig, A., & Song, D. (2003). Random key pre distribution schemes for sensor networks. *Proc. 19th Int. Conf. Data Eng.*, 197–213.

Chen, H., Meng, C., Shan, Z., Fu, Z., & Bhargava, B. K. (2019). A novel low-rate denial of service attack detection approach in ZigBee wireless sensor network by combining Hilbert–Huang transformation and trust evaluation. *IEEE Access: Practical Innovations, Open Solutions*, 7, 32853–32866.

Chen, Y., Liu, W., Niu, Z., Feng, Z., Hu, Q., & Jiang, T. (2020). Pervasive intelligent endogenous 6G wireless systems: Prospects, theories and key technologies. *Digital Communications and Networks*, 6(3), 312–320. doi:10.1016/j.dcan.2020.07.002

Cheng, Y., & Agrawal, D. (2005). Efficient pairwise key establishment and management in static wireless sensor networks. *Proc. 2nd IEEE Int. Conf. Mobile Ad Hoc Sensor Syst.*, 7.

Clincy, V., & Shahriar, H. (2019). IoT malware analysis. *Proc. IEEE 43rd Annu. Comput. Softw. Appl. Conf. (COMPSAC)*, *1*, 920–921.

Costanzo & Masotti. (2017). Energizing 5G. *IEEE Microwave Magazine*.

Das, A. K., & Zeadally, S. (2019). Data security in the smart grid environment. In Pathways to a Smarter Power System. Academic.

Devices & Systems. IoT Tech Expo. (2019). *Unlocking IoT Data With 5G and AI*. Available: https://innovate.ieee.org/innovation-spotlight/5g-iot-ai/

Dongand, Q., & Liu, D. (2007). Using auxiliary sensors for pairwise key establishment in WSN. *Proc. IFIP Int. Conf. Netw. (Networking)*, 251–262.

Du, W., Deng, J., Han, Y. S., Chen, S., & Varshney, P. K. (2004). A key management scheme for wireless sensor networks using deployment knowledge. *Proc. 23rd Conf. IEEE Commun. Soc. (Infocom)*, *1*, 586 597.

Du, W., Deng, J., Han, Y. S., & Varshney, P. K. (2003). A pairwise key pre distribution scheme for wireless sensor networks. *Proc. 10th ACM Conf. Comput. Commun. Secur.(CCS)*, 42–51.

Ericsson. (2019). *What is 5G?* Available: https://www.ericsson.com/en/5g

Eschenauer, L., & Gligor, V. D. (2002). A key management scheme for distributed sensor networks. *Proc. 9th ACM Conf. Comput. Commun. Secur.*, 41–47.

Fuentes, M., & Carcel, J. L. (2018, June). 5G New Radio Evaluation against IMT-2020 Key Performance Indicators. *IEEE Access: Practical Innovations, Open Solutions*, *8*, 110880–110896. https://doi.org/10.1109/ACCESS.2020.3001641

Gandotra & Jha. (2017). A survey on green communication and security challenges in 5G wireless communication networks. *Journal of Network and Computer Applications*, *96*, 39–61.

Granjal, Monteiro, & Sa Silva. (2015). Security for the Internet of Things: A survey of existing protocols and open research issues. *IEEE Commun. Surveys Tuts.*, *17*(3), 1294–1312.

He, D., Kumar, N., Khan, M. K., Wang, L., & Shen, J. (2018). Efficient privacy aware authentication scheme for mobile cloud computing services. *IEEE Systems Journal*, *12*(2), 1621–1631.

Hou, J., Qu, L., & Shi, W. (2019). A survey on Internet of Things security from data perspectives. *Computer Networks*, *48*, 295–306.

How to Avoid the Dreaded Computer Virus. (n.d.). Available: http://www. magellansolutions.co.uk/malware.html

How to Avoid the Dreaded Computer Virus. (n.d.). Available: http://www. magellansolutions.co.uk/malware.html

Hussain, R., & Zeadally, S. (2019). Autonomous cars: Research results, issues, and future challenges, *IEEE Commun. Surveys Tuts.*, *21*(2), 1275–1313.

Imran, M. A., Sambo, Y. A., & Abbasi, Q. H. (2019). Evolution of vehicular communications within the context of 5G systems. In *Enabling 5G Communication Systems to Support Vertical Industries* (pp. 103–126). IEEE. doi:10.1002/9781119515579.ch5

Jan, S. U., Ahmed, S., Shakhov, V., & Koo, I. (2019). Toward a lightweight intrusion detection system for the Internet of Things. *IEEE Access: Practical Innovations, Open Solutions*, *7*, 42450–42471.

Jo, K., & Sunwoo, M. (2014). Generation of a precise roadway map for autonomous cars. *IEEE Trans. Intell. Transp.*, *15*(3), 925–937.

Kamaldeep, M. M., & Dutta, M. (2017). Contiki-based mitigation of UDP flooding attacks in the Internet of Things. *Proc. Int. Conf. Comput., Commun. Automat. (ICCCA)*, 1296–1300. doi: 10.1109/CCAA.2017.8229997

Kao, Y.-W., Huang, K.-Y., Gu, H.-Z., & Yuan, S.-M. (2013). UCloud: A usercentric key management scheme for cloud data protection. *IET Information Security*, *7*(2), 144–154.

Khan, M. A., & Salah, K. (2018). IoT security: Review, block chain solutions, and open challenges. *Future Generation Computer Systems*, *82*, 395–411.

Khan, R., Kumar, P., Jayakody, D. N. K., & Liyanage, M. (2020). A survey on security and privacy of 5G technologies: Potential solutions, recent advancements, and future directions. *IEEE Commun.*, *22*(1), 196–248. doi:10.1109/COMST.2019.2933899

Kumar, R., Zhang, X., Wang, W., Khan, R. U., Kumar, J., & Sharif, A. (2019). A multimodal malware detection technique for Android IoT devices using various features. *IEEE Access: Practical Innovations, Open Solutions*, *7*, 64411–64430.

Lee, H. (2015). *Concept and Characteristics of 5G Mobile Communication Systems*. Available: https://www.netmanias.com/en/post/blog/7109/5g-iot/concept-andcharacteristics-of-5g-mobile-communication-systems-1

Li, J., Chen, X., Li, M., Li, J., Lee, P. P. C., & Lou, W. (2014). Secure deduplication with efficient and reliable convergent key management. *IEEE Transactions on Parallel and Distributed Systems*, 25(6), 1615–1625.

Limitless Intelligent ConnectivityM. W. C. (n.d.). Available: https://www.mwcbarcelona.com/

Liu, D., Ning, P., & Du, W. (n.d.). Group-based key pre-distribution in wireless sensor networks. *Proc. ACM Workshop Wireless Secur. (WiSe),* 1–14.

Liu, D., Ning, P., & Li, R. (2005). Establishing pairwise keys in distributed sensor networks. *ACM Transactions on Information and System Security*, 8(1), 41–77.

Liu, Z., Zhang, L., Ni, Q., Chen, J., Wang, R., Li, Y., & He, Y. (2019). An integrated architecture for IoT malware analysis and detection. In B. Li, M. Yang, H. Yuan, & Z. Yan (Eds.), *IoT as a Service* (pp. 127–137). Springer.

Messerges, T. S., Dabbish, E. A., & Sloan, R. H. (2002). Examining smartcard security under the threat of power analysis attacks. *IEEE Transactions on Computers*, 51(5), 541–552.

Nokia. (2016). *LTE Evolution for IoT Connectivity*. Nokia, Tech. Rep.

Northeast Now. China: Shanghai's Hongkou District Becomes First With 5G Network in World. (n.d.). Available: https://nenow.in/neighbour/china-shanghais-hongkou-district-becomesfirst-with-5g-network-in-world.html

Perrig, A., Szewczyk, R., Tygar, J. D., Wen, V., & Culler, D. E. (2002). SPINS: Security protocols for sensor networks. *Wireless Networks*, 8(5), 521–534.

Pham, C., Nguyen, T., & Dang, T. (2019). *Resource-Constrained IoT Authentication Protocol: An ECC-Based Hybrid Scheme for Device-to-Server and Device-to-Device Communications*. doi:10.1007/978-3-030-35653-8_30

Qualcomm. (2019). *Everything You Need to Know About 5G*. Available: https://www.qualcomm.com/invention/5

RPMA. (2016). *RPMA Technology for the Internet of Things*. Ingenu, Tech. Rep.

Shafi, M., Molisch, A. F., Smith, P. J., Haustein, T., Zhu, P., De Silva, P., Tufvesson, F., Benjebbour, A., & Wunder, G. (2017). 5G: A tutorial overview of standards, trials, challenges, deployment, and practice. *IEEE Journal on Selected Areas in Communications*, *35*(6), 1201–1221.

SigFox. (n.d.). www.sigfox.com

Su, J., Vasconcellos, V. D., Prasad, S., Daniele, S., Feng, Y., & Sakurai, K. (2018). Lightweight classification of IoT malware based on image recognition. *Proc. IEEE 42nd Annu. Comput. Softw. Appl. Conf. (COMPSAC)*, *2*, 664–669.

Taylor, L., Fici, G. P., & Hersent, O. (2011). Interconnecting Zigbee & M2M Networks. *ETSI M2M Workshop*, 1-18.

Tian, Z., Sun, Y., Su, S., Li, M., Du, X., & Guizani, M. (2019). *Automated attack and defense framework for 5G security on physical and logical layers.* Available: https://arxiv.org/abs/1902.04009

https://medium.com/illumination/5g-fifth-generation-of-mobile-networks-part-1f32d7f003686

Tysowski & Hasan. (n.d.). Hybrid attribute-and re-encryption based key management for secure and scalable mobile applications in clouds. *IEEE Trans. Cloud Comput.*, *1*(2), 172–186.

Vangelista, L., Zanella, A., & Zorzi, M. (2015). Long-range IoT technologies: The dawn of LoRaTM. In *Future Access Enablers of Ubiquitous and Intelligent Infrastructures* (pp. 51–58). Springer.

Wang, D., Cheng, H., He, D., & Wang, P. (2018). On the challenges in designing identity-based privacy-preserving authentication schemes for mobile devices. *IEEE Systems Journal*, *12*(1), 916–925.

Wazid, Das, Hussain, & Succi, & Rodrigues. (2019). Authentication in cloud-driven IoT-based big data environment: Survey and outlook. *Journal of Systems Architecture*, *97*, 185–196.

Wazid, M., Bagga, P., Das, A. K., Shetty, S., Rodrigues, J. J. P. C., & Park, Y. (2019). AKM-IoV: Authenticated key management protocol in fog computing based Internet of vehicles deployment. *IEEE Internet Things J.*, *6*(5), 8804–8817.

Wazid, M., & Das, A. K. (2016). An efficient hybrid anomaly detection scheme using K-means clustering for wireless sensor networks. *Wireless Personal Communications*, *90*(4), 1971–2000.

Wazid, M., & Das, A. K. (2017). A secure group-based blackhole node detection scheme for hierarchical wireless sensor networks. *Wireless Personal Communications*, *94*(3), 1165–1191.

Wazid, M., Das, A. K., Hussain, R., Succi, G., & Rodrigues, J. J. P. C. (2019). Authentication in cloud-driven IoT-based big data environment: Survey and outlook. *Journal of Systems Architecture*, *97*, 185–196.

Wazid, M., Das, A. K., Kumar, N., Conti, M., & Vasilakos, A. V. (2018). A novel authentication and key agreement scheme for implantable medical devices deployment. *IEEE Journal of Biomedical and Health Informatics*, *22*(4), 1299–1309.

Wazid, M., Das, A. K., Kumar, N., Vasilakos, A. V., & Rodrigues, J. J. P. C. (2019). Design and analysis of secure lightweight remote user authentication and key agreement scheme in Internet of drones deployment. *IEEE Internet Things J.*, *6*(2), 3572–3584.

Wazid, M., Das, A. K., Kumari, S., & Khan, M. K. (2016). Design of sinkhole node detection mechanism for hierarchical wireless sensor networks. *Security and Communication Networks*, *9*(17), 4596–4614.

Wazid, M., Das, A. K., & Lee, J.-H. (2019). User authentication in a tactile Internet based remote surgery environment: Security issues, challenges, and future research directions. *Pervasive and Mobile Computing*, *54*, 71–85.

Wazid, M., Das, A. K., Odelu, V., Kumar, N., & Susilo, W. (2020). Secure remote user authenticated key establishment protocol for smart home environment. *IEEE Transactions on Dependable and Secure Computing*, *17*(2), 391–406.

Wazid, M., Das, A. K., Shetty, S., Gope, P., & Rodrigues, J. J. P. C. (2020). Security in 5G-Enabled Internet of Things Communication: Issues, Challenges and Future Research Roadmap. *IEEE Access: Practical Innovations, Open Solutions*, *8*, 1–25. https://doi.org/10.1109/ACCESS.2020.3047895

Wazid, M., Das, A. K., & Vasilakos, A. V. (2018). Authenticated key management protocol for cloud-assisted body area sensor networks. *Journal of Network and Computer Applications*, *123*, 112–126.

Wazid, M., Dsouza, P. R., Das, A. K., Bhat, V. K., Kumar, N., & Rodrigues, J. J. P. C. (2019). RAD-EI: A routing attack detection scheme for edge-based Internet of Things environment. *International Journal of Communication Systems*, *32*(15), 4024. doi:10.1002/dac.4024

Wu, B., Xu, K., Li, Q., Liu, Z., Hu, Y.-C., Zhang, Z., Du, X., Liu, B., & Ren, S. (2019). Decentralized and automated incentives for distributed IoT system detection. *Proc. IEEE 39th Int. Conf. Distrib. Comput. Syst. (ICDCS)*, 1106–1116.

Wu, L., Wang, J., Choo, K.-K.-R., & He, D. (2019). Secure key agreement and key protection for mobile device user authentication. *IEEE Transactions on Information Forensics and Security*, *14*(2), 319–330.

Yang, Y., Wu, L., Yin, G., Li, L., & Zhao, H. (2017). A survey on security and privacy issues in Internet-of-Things. *IEEE Internet Things J.*, *4*(5), 1250–1258.

Yu, J., Ren, K., Wang, C., & Varadharajan, V. (2015). Enabling cloud storage auditing with key-exposure resistance. *IEEE Transactions on Information Forensics and Security*, *10*(6), 1167–1179.

Zhou, W., Jia, Y., Peng, A., Zhang, Y., & Liu, P. (2019). The effect of IoT new features on security and privacy: New threats, existing solutions, and challenges yet to be solved. *IEEE Internet Things J.*, *6*(2), 1606–1616.

Zhu, S., Setia, S., & Jajodia, S. (2006). LEAP+: Efficient security mechanisms for large-scale distributed sensor networks. *ACM Transactions on Sensor Networks*, *2*(4), 500–528.

KEY TERMS AND DEFINITIONS

5G (Fifth Generation): Designed to improve 4G boundaries and performance. It has high connectivity and a fast data transfer rate with minimal latency. It facilitates communication between device and device while consuming less power and provides better telephony. 5G data transfer faster than 4G, which can reach a maximum of 35.46 Gbps. Technologies running on the theme of Multi Input and Multiple Output (MIMO) and Millimeter Wave Communications.

Attacks: It is a data security threat that involves an attempt to obtain, alter, destroy, remove, implant, or reveal data without authorized access or permission.

Authentication: It is the process of recognizing a user's identity which verifying the identity of user or information.

Internet of Things (IoT): The internet of things is defined as network of interconnected with devices, system of interrelated computing devices machinery, people or animal, items with unique identification to send data over network.

Mobile Communication: It allows us to communicate with others in various locations without use of any physical connection.

Networking: Number of connected devices which are sharing and acquiring information between interconnected devices.

Privacy: Privacy is the protection of computer system from theft or devices.

Security: The security that prevents unauthorized access to computers, network, and data.

Chapter 2
5th Generation Security Threats and Responses

Sumit Dhariwal
Manipal University Jaipur, Jaipur, India

Avani Sharma
Manipal University Jaipur, Jaipur, India

ABSTRACT

5G could help with the extremely dependable and cost-effective networking of a huge number of devices (e.g., internet of things [IoT]), as well as universal broadband access and high user mobility. The current technical enablers for 5G are cloud computing, software-defined networking (SDN), and service-oriented virtualization (NFV). However, these technologies offer security issues in addition to raising concerns about user privacy. In this chapter, the authors give an outline of the security difficulties that these technologies face, as well as the privacy rules that apply to 5G. They also offer security remedies to these problems, as well as research directions for dependable 5G frameworks.

INTRODUCTION

The Telecommunication Standardization Union (ITU) has named the International Leading Cellular standard as the worldwide requirement per the 5G wireless communication. IMT-Advanced is the 5th generation of 5G network technologies used in a cellular network. The international telecommunication union's radio communications area has said that 5G requires (Agiwal et al., 2016) I have always had the following characteristics: high mobility; ultra-reliability and low latency

DOI: 10.4018/978-1-6684-3921-0.ch002

(1 Ms); andhigh peak data throughput of 10–20 Gb. Many 5G services are used to provide and work to construct systems and the matches need to be helpful and provide the 5g working environment for particularly greater speeds, but they lack wireless and wired network convergence. The Wi-Communication (Wi-Com) architecture created by the University of Cambridge, on the other hand, provides for heterogeneous networking. A fast core network and a sluggish periphery network are included (Alliance, 2015).

With us, the system is a combination of optical networks and peripheral communications, and it makes use of wireless technologies similar to the 5G system (3GPP, 2017), which is a significant part of the network and that network should be the main. (Security, 2013) (Kulkarni et al., 2016) (Vikas et al., 2014) Security issues in 4G operations have been carefully explored. Furthermore, for 5G heterogeneous mobile networks, there is presently no effective security protection. Even though Wi-Com contains a security solution with a multi-layered security system, investigations have revealed that many security risks might cause service disruptions and data loss. There are present and new perceived security concerns in 5Mobile communications, according to a study (La Polla et al., 2012).

Researchers from ETH_zurich university of Lorraine, and the other university like the university of dundee discovered that thieves might retrieve 5 G-related of communications networks and steal data due to multiple security flaws. This is due in part to "unspecified security goals" and "lack of precision," according to a news statement from the group (Suo et al., 2013). As a result, 5G heterogeneous network requirements may be divided into two main categories: Mobile devices and operating company networks are both connected. Furthermore, certain transport security specifications, and must be useful and considered as ensuring the integrity of devices, discretion; ensuring controlled access to data; and preventing mobile devices from being stolen or tampered with, thus preventing data from being tampered with because it appears that 5G implementations require a subscriber's whole security architecture, we developed a new policy for a security management system to identifies whether attackers are using mobile devices in Wi-com situations. The model utilizes ITU_T guideline M.3400 to deal with data breaches, whether it has been done or not:

1. An omniscient omnipotent (IA) approach for identifying malicious activity in edge devices.
2. An identity cell that interacts with the Wi-Com system's classifiers.
3. A regulatory virtuous loop generated by an end-user device is depicted as the personality cell.

BACKGROUND

The 5G Network

Communication and information Applied science has given rise to global innovations. The increasing ability to transfer and process information quickly, e-commerce, social engagement, content delivery systems, e, and m-learning, as well as video & audio including conferencing, are just a few examples is transforming society in a variety of ways. Industry and trade have historically relied heavily on technological advancement, it's getting harder to disregard what's expected in the next-generation wireless network communications to fulfill growing customer demands (Chonka & Abawajy, 2012).

In meeting the increasing demands of users (LTE), next generation 5G wireless communications must achieve much higher data rates, lower latency, increased base station capacity, and better QoS than current 4G networks. Is. The following are the key requirements According to key industries, researchers, and suppliers, the following are the requirements for next-generation 5G systems:

- Realistic networks with up to 10 Gbps data speeds (a 10-fold increase compared to LTE networks) (Sucasas et al., 2016) Formal euphemisms
- Greater bandwidth per capita than 4G (Ahmad et al., 2015)
- Increase the number of connected device subscribers in order to realize a more creative and presentational use for IoT (Fonseca et al., 2012)
- The round trip of one-millisecond latency used in ten times.
- Broad coverage ("anytime, anyplace" connection) — A 5G wireless network must have comprehensive coverage. Damn near-comprehensive coverage
- Brings down power requirements by around 90%.

Y-Com Architecture

A team of academics from Middlesex University's connectivity research organization, Cambridge university's computer Laboratory, and the research centre of Samsung and Deutsche Telekom created the Wi-Com framework. The architecture is designed to address emerging problems in heterogeneous networks. There are issues at the network, device, and application levels, among others. The framework has a layered approach and functions as a prototype in the same way that the program used in the OSI reference model used in the network model. (Park & Park, 2007b). In this paper, we present a security management system for the Y-Com framework.

Analysis of Y-Comm and 5G Networks

5G infrastructure may be accessible via several outdoor and externally connected system points and others including peer operators with the internet, and third-party technologies used to understand the system but due to its open nature. Because service providers use the same basic network architecture, if one provider is hacked, the entire network infrastructure is in danger (Strassner, 2004). Although the solutions are not complete, the Wi-Com research group has devised a security mechanism to resist such attacks. Arash and his colleagues concentrated their study on the security concerns that 4G networks face. Because he invented to find and discovered that the existing value and new upcoming security concerns were specific to 4G_technology, his solution to these challenges involved extending current security methods to 4G networks.

Policy Overview

This article explains how to resolve vulnerabilities in heterogeneous network security systems using a policy-based approach. 5G heterogeneous networks are complicated by the convergence of both wired and wireless systems and technology as well as the variety for the network technology. Researchers have been pushed to create acceptable network management approaches due to the difficulties involved. displaying management systems based on policies total assurance Such connection requires control measures. As described in (Asim et al., 2018; Neisse et al., 2008; Töyssy & Helenius, 2006; Twidle et al., 2009; Zhao et al., 2008; Zhou et al., 2012), there are various reasons for the present interest in designing governance software suite:

Obligation Policies

Limiting policies define what a system subject should do in the case of a given occurrence. Security policies are triggered by predefined events. Occasion Condition Action (ECA) (Vook et al., 2014) is the basis for the regulations. Security breaches are one example of how liability insurance might be employed. A variety of procedures are done to secure the network when a password is compromised. In this work, we employed such strategies to respond to pre-set security breaches. The duties are based on the guidelines of the International Telecommunication Union and a policy design language that permits attribution rules to function in wireless links, was employed in this investigation.

Figure 1. Key endpoint structure with network connectivity

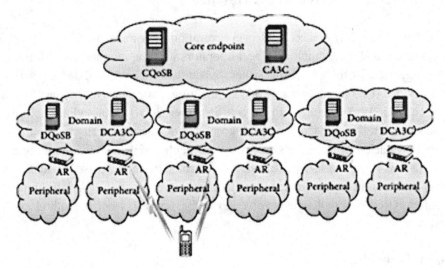

Policy System Selection

The ponder and PDL, with the XACML & LASCO, Tower, and the other Ponder2 are some regulatory mechanisms that have influenced the adoption of systems that are acceptable for the profession.

Each policy's benefits and disadvantages were evaluated to see if it was suitable for inclusion in a secure environment to create the heterogeneous network for 5g. While these policy systems cover the main policy types required with help for security for management, they are generally built for managing large, dispersed information systems, except for pondering, which is intended for smaller devices. are planned. are not acceptable. Ponder-2 is different from Ponder and PDL in that it is more versatile and customizable and capable of adapting to a wide range of network architectures and system software. Ponder-2 also offers Ponder Talk, a restructuring language that eliminates the need for the management system's author to be familiar with the low-level specifics of the different tools Ponder-2 requires in such an environment. required. Makes an excellent pick because it covers a wide range of topics. Different network technology plus home appliances

WIRELESS NETWORKS SECURITY

Wireless security is very useful for the security reason of complexity in the underlying networks, proprietary and perimeter-based security solutions that are difficult to

Table 1. Summary of security evaluation from One-G to Five-G

Security Challenges	Network Used	Security Mechanisms
No explicit security and privacy measures	One-G	Eavesdropping calls interception, and no privacy mechanism.
Authentication anonymity and encryption-based protection.	Two-G	Replica base station, With links of Radio and security, With one-way authentication and spamming work.
Adopted the two-g Mutual authentication scheme agreement and two-way authentication) were established to secure network access.	Three-G	Vulnerability of IP traffic, encryption, key protection, and roaming protection.
New encryption (EPS AKA) and trust methods were introduced, as well as encryption key protection and non-3GPP Partnership Project (3GPP) access security and integrity protection.	Four-G	Increased IP vehicle security, such as DOS overlay information security, base transceiver station reliability, and long-term key eavesdropping, also isn't appropriate for five-G confidentiality.

administer, and weaknesses in identity management, security of communication networks from 1-G to 4-G has been a tough challenge (Kim et al., 2018). Furthermore, Internet architecture inherits infrastructure-related issues, poses security risks, and is willing to try new things (Zheng et al., 2005). IEEE 802.11 security has increased exponentially with the introduction of mobile networks (Park & Park, 2007a). The most major change started with the invention of IP-based communication in wireless networks, which placed an increasing amount of Internet-based security problems onto wireless networks. As a result, in this paragraph, we provide an account of the changing security paradigm in wireless networks (summarised in Table 2) as they migrate from 1G to 4G or non-IP to Mac address Wi-Fi networks.

Security in Non-IP Networks

1G communications networks employed analogy signal amplification and were predominantly geared for IP telephony (Dierks & Rescorla, 2008). Advanced phone service (Santos et al., 2013), the most successful 1-G technology, was initially commercially implemented in 1983 by AT & T and Bell Labs. 1-G found it challenging to deliver effective security services due to the nature of analogy transmission. Knowledge of voice calls was not safe since this smartphone service did not employ encryption. As a means, eavesdropping, illegal access, cloning, and user privacy were all possible security issues (YuHunag et al., 2010)[29]. To increase the effectiveness of restricted frequency bands, digital mobile systems were developed (Jafarian et al., 2012) (Gember et al., 2012), and therefore, they have become much more productive in cellular telephones as parts of the cell connection.

Security in 3G

3G Compared to 2G networks, mobile communication networks may deliver significantly quicker data rates. New mobile network services like video calling and video streaming have also been made possible by 3G systems (Norrman et al., 2016).

The 3G standard suggested a more secure security structural design to solve the weaknesses of second-generation systems works. Three essential concepts of the third-generation security are specified in the 3GPP (Khan et al., 2013): (i) We talk about the third generation security will take over essential elements of 2G security, (ii) Third generation security will push the limits of second-generation work to gather and improve the security, and (iii) The Third generation security will be more specific outcome to create and privacy and security abut we add some essential feature Security features not present in second-generation mobile network but it will be added. Universal Mobile Telecommunications System (UMTS), a 3G cellular technology, was created and is maintained by 3GPP. TS33.102 (Khan et al., 2013) specifies five security aspects for the UMTS security architecture.

This is commonly referred to as release 99. This set of UMTS security features ensures that UE has secure access to 3G services and protects radio access links from attacks. (Cai et al., 2015). The UMTS Authentication and Key Agreement (AKA) protocol aim to be as compatible as possible with GSM. However, UMTS AKA achieves additional protocol goals, such as network mutual authentication and agreement on integrity keys, among others. Unlike GSM, which supports only one-way authentication, UMTS supports bilateral authentication, which eliminates the risk of a false base station. The Access Security feature protects the user's identity and ensures that the user cannot listen over the radio access link.

User location privacy and the ability to be untracked should always be supported by key exchange confidentially. The client is recognized by a temporary or adjacent permanent encrypted identity to satisfy these functions. Conversely, the identity of the user should also not be revealed as known for long periods of time, and any data that may betray the personality of the user should be encrypted (Jaeger, 2015).

Security in 4G

The International Telecommunication Union - Radio Communications Area (ITU-R) 4G standard (Lauer & Kuntze, 2016) is met by 5G Network Release 10, often known as LTE-Advanced signal strength. The LTE-A network is made up of three distinct parts.

Advanced Packet Core and Advanced-Universal Terrestrial Access Network are two protocols that have evolved in recent years (E-UTRAN). EPC is an IP and packet-switched backbone network that is fully operational.) LTE-A systems may connect to non-3GPP access networks. Machine-type communication (MTC), home anodes or femtocells, and relay nodes were all introduced as novel entities and applications. 3GPP has established a comparable set of defence characteristics for LTE-A signal strength:

1. Gain access to management (Developed Universal Terrestrial Radio Access Network.)
2. System territory security,
3. Customer territory defence,
4. User area defense user domain safety, user domain safety, user domain safety.
5. It will help protect the user domain safety,
6. as well as application domain security
7. Security configuration and visibility (Bikos & Sklavos, 2012)

Although each element has been greatly upgraded to protect LTE systems, Furthermore, totally new protection techniques for MTC (Zhao et al., 2017) (Ahmad et al., 2017), home ENB (Alliance, 2015), and relay nodes (Andrews et al., 2014) have been established.

5G-SECURITY

History In a highly dependable and cost-effective way, 5G will also provide widespread internet, stimulate IoT connectivity to many devices, and amuse high-mobility people and gadgets (Lu et al., 2014). 4G's growth of IP-based connectivity has facilitated the creation of future commercial potential; though, the fifth generation is seen as a new ecosystem that connects practically all areas of society, including vehicles, electronics, health care, commerce, and company.

This evolution will present new risks and security weaknesses, offering a serious challenge to both existing and future networks (Elijah et al., 2015). By linking a power system, for example, the fifth generation will generate essential energy infrastructure, and security 6 breaches in such critical infrastructure might have disastrous effects on both the infrastructure and the society that 5G serves. As a result, the security of, the fifth generation and its connected systems must be considered from the start of the design process. To expound on the potential repercussions of 5G, below is an overview of its design.

Overview of 5G

The principles of design with new-found services and gadgets, as well as new user constraints for low latency, high throughput, ubiquitous analysis, and new 5G design principles (Ahmad et al., 2016), are required. NGMN's 5G design methods, depicted in Figure 2, emphasize the importance of highly elastic and robust systems. Spectrum resource performance, cost-efficient compact deployment, effective coordination, interference suppression, and dynamic radio configuration are all requirements for radio modules. Beyond radio, networks have different requirements that are more focused on incorporating fundamentally new technologies. SDN and NFV, for example, would be used in a typical composable to allow the client to segregate and regulate the location of aircraft and dynamic network functions (Costa-Requena et al., 2015). Its goal is to eliminate old networking and replace it with new interfaces for core and edge devices.

General Idea of 5G Safety Planning

The security architecture is used according to the ITU-T (Namal et al., 2016) and splits protection aspects logically into the different types of shapes and sizes of the sections. It allows for a methodical approach to the total security of new services, which supports the creation of urban security solutions and the evaluation of current network security. The 5G security architecture is organized into domains in the current 3GPP technical standard release (Release 15) (Ahmad, 2020). Figure 3 shows the network security excluding domain (VI), which contains the following major domains.

- Protecting Network Access (I): UE can safely authenticate and access network services thanks to a set of security measures. Gain access to protection and it includes the protection of 3gpp and non-3GPP access technologies, as well as the providing and it gives to the next level security for context from the SN and according to the UE.
- Internet backbone Sector Privacy (II): The collection of security measures and it enables signals to the network nodes as well as the safe sharing of user aircraft data.
- user Access Domains Security (III): The security of the user's domain is discussed here. It has security measures that allow users to access the UE in a secure manner.
- Map Function Theoretical and practical: The application domain's security. It has security capabilities that allow applications (both user and provider domains) to safely send and receive messages.

- Service Premises Architecture (SBA): Mixed security features for Registration, detection, and authorization of network elements, as well as service-based interface security
- Range and Order Capability: Contains security elements for the capabilities that notify consumers when security features are active.

Table 2. 5G design principles

Radio	Network	Operation & Management
Spectrum efficiency	Create a common composable core	Simple operation and management. Automation self-healing
Cost-effective	Minimize entities and function control unit function	Problem monitoring
Coordinate and close interference	RAT core	Collaborative management
Support dynamic radio topology	Minimize legacy	Integrated OM functionality
		Network cloud orchestration
Flexibility Functions	**Support new value creation**	**Build in security**
Network slicing	Big data and contact awareness	HetNets
Function variance	Radio networks API	Privacy
Flexible function	Facilitate XaaS	Identity protection
Leverage NFV/SDN		
State function		
Graceful degradation		

5G SECURITY: AGENERAL OVERVIEW

NGMN (Another Generation of Wireless Communication Networks) produced 5G proposals based on gaps in existing network designs and security measures that are either being deployed or are not already in use. This proposal stresses cautious statements, such as 5G's infancy, which is riddled with mysteries and a shortage of well- Unfamiliar end-to-end and subsystem architectures and design concepts. This proposal emphasizes access network security issues as well as cyber assaults on applications and network equipment, as demonstrated in Figure 4.

Strobe traffic on the network: This same frequency of end hardware in 5G is expected to significantly boost, resulting in significant changes in network traffic patterns, either accidentally or deliberately. As a response, 5Mobile operators must be

Figure 2. A high-level overview of the security architecture.

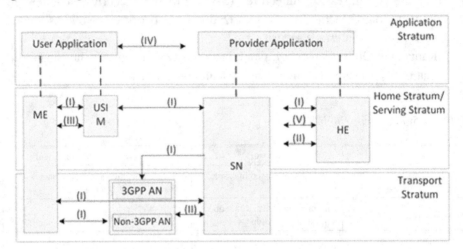

strong enough to withstand massive traffic variations while maintaining acceptable levels of performance.

In previous network topologies, including 4G, radio interface encryption keys were incorporated into home networks and transported over unprotected connections over visiting networks, exposing the keys. As a result, it's advisable to secure the keys beforehand, rather than delivering them via an unsecured channel like SS7/DIAMETER.

Dissatisfaction with Mobility Management: Some signals are protected by 3G and 4G systems, however, ethical integrity is not protected by decrypting the customer data plane. As a result, security should be limited to the transport or application layer, which extends beyond mobile networks. Despite this, application-level end-to-end security might take a significant amount of labor for the delivery of data in packet headers and greetings. As a result, resource-constrained IoT devices, network-level security, or latency-sensitive IoT devices are becoming increasingly important.

Information infrastructure mandated by law: Security infrastructure might suffer from system limitations (such as latency), forcing the deployment of alternative security methods. Regrettably, such limits undercut system-level security assumptions and cannot be totally removed. In multi-operator circumstances, the problem is worsened when an operator has afflicted losses due to insufficient preventive measures. As a result, following a comprehensive examination to identify the most serious security concerns, it is highly recommended that some degree of security be enforced in 5G.

Figure 3. Threats to 5G network security

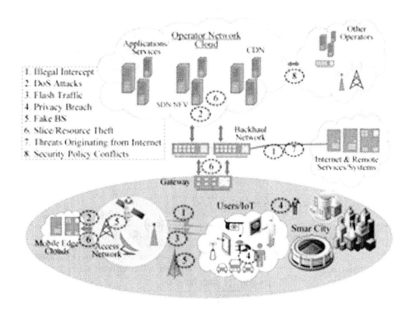

Consistent security rules at the consumer level are required: When a user switches from one operator network to another, user security measures must be maintained. It's possible that security services aren't updated as frequently as they should be.

INFRASTRUCTURE DENIAL-OF-SERVICE ATTACKS

Rather than a DoS, Assaults on key infrastructure including energy, health, transportation, and telecommunications networks can be disrupted by distributed DoS (DDoS) attacks. Attackers are often intended to draw down the physical hardware capacities of the attacking device. The possibility of assaults from widely distributed and internationally distributed machines will exacerbate the problem (compromised IoT). As a result, the network's security must be strengthened. The multiplicity of devices and services in 5G complicates security. We explain potential security solutions, including methods and suggestions from various regulatory and standards authorities, in the following sections, even at a high level.

5G NETWORK SECURITY ISSUES AND SOLUTIONS

Security in the network design is presented at three levels:

Figure 4. 5G Security Vision and Goals

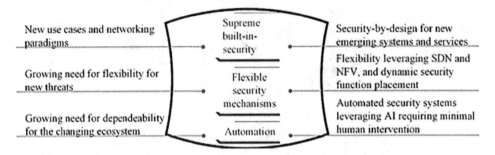

Table 3. Security dimension defend by ITU-T

S.No	Security Dimension	Brief Explanation
1	Gain access to control	In the realms of environmental security and information security, access control (AC) is the selective restriction of access to a facility or other resource.
2	Authentication	Another authentication mechanism available in 5G is EAP-AKA. It's also a challenge-and-response system that relies on a The UE and its home network to share a cryptographic key. It has the same degree of security as 5G-AKA, including user and network mutual authentication.
3	Non-Reputation	The conviction that nothing can be denied is referred to as a disclaimer. In the information sector, the word "no disclaimer" is often used.
4	Data-Confidentiality	Unless the information must be kept secret due to laws, rules, or contracts, if the data may only be used or published if certain criteria are met, if somehow the data is vulnerable and would have negative consequences if released, And if it can be useful to those who aren't supposed to have access to it, that's even better (for example, hackers)
5	Communication security	Communication and information protection, in 5G/6G networks, multimedia secure communications are possible. Forensics of media in 5G/6G networks, 5G/6G network cross-media security, communication protocol security in 5G/6G networks, in 5G/6G, live content security detection is possible. Developing multimedia communication systems that are secure, Management of security risks, and multimedia data evaluation in 5G/6G networks.
6	Data integrity	Throughout the life cycle of data, information governance is an important aspect of the design, development, and maintenance of data correctness and consistency.
7	Availability	Verizon and other firms are currently offering 5G fixed wireless broadband internet, as are Verizon, AT&T, and T-Mobile, as well as a few smaller companies.
8	Privacy	The entities involved in 5G deployment can jointly take a proactive approach to by incorporating privacy protocols into the core architecture of 5G networks.

Figure 5. A high-level architecture of a 5G network is presented.

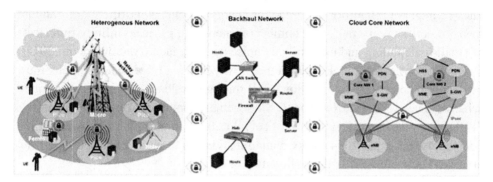

1. Reach the network,
2. the backhaul network, and
3. the main network, to correctly assess the security viewpoint of the whole network in a methodical manner.

To be explicit, we first identified the security issues (which are also shown in Table III) before discussing potential remedies.

5G TECHNOLOGY SECURITY IS CRITICAL

The security issues of 5G may be properly defined by evaluating the primary supporting technologies. In comparison to earlier generations, cloud storage ideas such as MIMO arrays, SDN, NFV, and multi-access network virtualization are the major enabling and disruptive technologies (MEC). Because SDN, NFV, and cloud computing have all been utilized in non-wireless networks, there is an abundance of security literature. We'll talk about their security issues and solutions in this part, with an emphasis on their utilization in 5G networks. The first topic is security problems (as shown in Table VI), accompanied by software applications or endeavors.

Massive MIMO Security

Massive MIMO Security Issues

Massive MIMO is one of 5G's most promising and disruptive technologies. Large-scale MIMO is based on the notion of equipping a base station with a huge proportion of antenna sections large enough to support many user terminals on the same radio spectrum (3GPP, 2017).

A large percentage of antenna arrays are used in a variety of ways to increase data rates or improve reliability, coverage, or energy efficiency. Furthermore, random matrix theory suggests that as the number of antennas approaches infinity, the effects of small-scale fading and uncorrelated noise begin to fade (Andrews et al., 2014). Despite its promises, to take advantage of MIMO on a large scale, stations must estimate Channel State Information (CSI) through feedback or channel reciprocity schemes. The use of non-orthogonal pilot schemes for multicell time division duplex (TDD) networks introduces the concept of pilot contamination due to coherence time constraints (Lu et al., 2014). Pilot contamination has a massive impact on MIMO at large. These are called pilot pollutants.

SDN's Security

The network control plane is separated from the forwarding plane in SDN, and network control is centralized in a software-based network control platform. Logically controlled software network control functions interface with forwarding devices through programmable APIs. It simplifies network control, administration, and operation while also speeding up the development and implementation of in-network facilities. As a result, academia and industry have been focusing on the use of SDN ideas in wireless networks. As a result, various ideas for SDN-based wireless networks have been made (Elijah et al., 2015)–(Namal et al., 2016). As depicted in Figure 6, the SDN architecture consists of three functional levels with interconnections between them. With OpenFlow Application, OpenFlow Controller, and OpenFlow Switch, OpenFlow is the first practical implementation of SDN that follows the three-tier architecture of SDN.

NEW DIMENSIONS IN THE SECURITY OF FUTURE NETWORKS

Another security paradigm change was required due to the substantial expansion in the kind and quantity of IoT devices, the notion of networked smart societies, or the development of a genuinely new gadget-free society (Dierks & Rescorla, 2008). Advances in digital paradigms, such as quantum computing, will necessitate the development of new robust security designs that leverage powerful computation to improve network security. With super-fast or periodic shared compute, cracking cryptographic hash functions might become a reality.

The core premise is that security solutions should match the diversity and growth of computer and communications technologies. In this part, we provide a complete overview of the next generation of wireless networks (XG) (Figure 9) and describe how to improve their security using cutting-edge technology concepts that are now a

Figure 6: Architecture for SDN

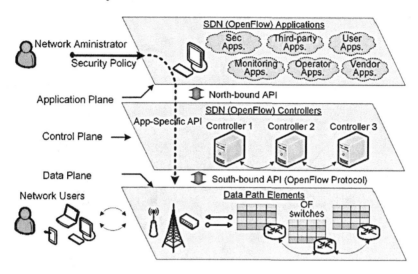

hot research issue. For example, super-fast security mechanisms at various network nodes and places, such as network edge or fog nodes, security automation and self-healing, all of which will necessitate AI development, as well as the safe provisioning of surveillance services. This will need the use of blockchain technology.

Future Privacy Measures

5G as well as similar connections will present greater privacy problems than ever before. These issues exist not only from the standpoint of consumers, but also from the standpoint of telecommunications companies, network operators, vendors, and CSPs to assure effective network implementation. Blockchain and edge computing/cloud are enabling technologies that will be incorporated into future networks in 5G and beyond, and they are required to fulfill some standards and guidelines. However, because the future network will be made up of numerous stockholders with various corporate interests, it does not completely secure your privacy. Consequently, preserving secrecy is a difficult effort for everyone involved in the network. As a result, it's critical for regulatory agencies to create privacy laws designed to protect customers' anonymity while simultaneously safeguarding the interest of all stakeholders involved. Additionally, new security procedures can be coupled with old cryptographic approaches. To guarantee that only legitimate entities are authenticated and authorized. Future privacy solutions will rely heavily on approaches like privacy-by-design and personalized privacy. However, because

Figure 7. A look at future municipal communication networks' security systems (highlight AI, SDN, and NFV in the picture).

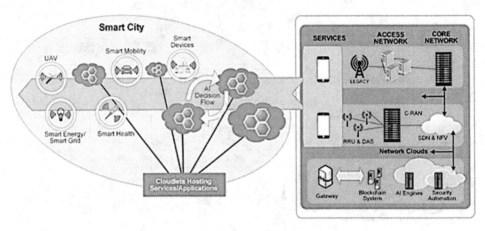

privacy requires an awareness of the context, AI-based privacy solutions will be critical for future communications.

CONCLUSION

To meet the difficulties of connection, flexibility, and cost-scale, 5G will leverage mobile cloud, SDN, and NFV. These technologies, while their numerous advantages are not without security threats. As a result, we've emphasized the primary security problems in this article, which, if not handled effectively, would make 5G much riskier. We also spoke about security procedures and remedies to these problems. However, the security threat vectors have yet to be completely realized due to the limited standalone and integrated implementation of these technologies in 5G. Similarly, when more consumer devices, such as IoT, become linked and 5G introduces a slew of new services, the threats to communication security and privacy will become more obvious. In summary, when new 5G technology and services are implemented, new sorts of concerns and vulnerabilities are expected to emerge. Taking these problems into account from the beginning of the design process to the result, on the other hand, will lessen the risk of future security and privacy vulnerabilities.

REFERENCES

3GPP. (2017). *SA3-Security. The Third Generation Partnership Project (3GPP)*. Available: https://www.3gpp.org/Specifications-groups/sa-plenary/54-sa3-security

Agiwal, M., Roy, A., & Saxena, N. (2016). Next generation 5G wireless networks: A comprehensive survey. *IEEE Communications Surveys and Tutorials*, *18*(3), 1617–1655. doi:10.1109/COMST.2016.2532458

Ahmad, I. (2020). *Improving software defined cognitive and secure networking*. arXiv preprint arXiv:2007.05296.

Ahmad, I., Liyanage, M., Namal, S., Ylianttila, M., Gurtov, A., Eckert, M., . . . Ulas, A. (2016, January). New concepts for traffic, resource and mobility management in software-defined mobile networks. In *2016 12th Annual conference on wireless on-demand network systems and services (WONS)* (pp. 1-8). IEEE.

Ahmad, I., Liyanage, M., Ylianttila, M., & Gurtov, A. (2017, June). Analysis of deployment challenges of host identity protocol. In *2017 European Conference on Networks and Communications (EuCNC)* (pp. 1-6). IEEE. 10.1109/EuCNC.2017.7980675

Ahmad, I., Namal, S., Ylianttila, M., & Gurtov, A. (2015). Security in software defined networks: A survey. *IEEE Communications Surveys and Tutorials*, *17*(4), 2317–2346. doi:10.1109/COMST.2015.2474118

Alliance, N. G. M. N. (2015). 5G white paper. *Next generation mobile networks, white paper, 1*(2015).

Andrews, J. G., Buzzi, S., Choi, W., Hanly, S. V., Lozano, A., Soong, A. C., & Zhang, J. C. (2014). What will 5G be? *IEEE Journal on Selected Areas in Communications*, *32*(6), 1065–1082. doi:10.1109/JSAC.2014.2328098

Asim, M., Yautsiukhin, A., Brucker, A. D., Baker, T., Shi, Q., & Lempereur, B. (2018). Security policy monitoring of BPMN-based service compositions. *Journal of Software: Evolution and Process*, *30*(9), e1944.

Bikos, A. N., & Sklavos, N. (2012). LTE/SAE security issues on 4G wireless networks. *IEEE Security and Privacy*, *11*(2), 55–62. doi:10.1109/MSP.2012.136

Cai, Y., Yu, F. R., & Bu, S. (2015). Dynamic operations of cloud radio access networks (C-RAN) for mobile cloud computing systems. *IEEE Transactions on Vehicular Technology*, *65*(3), 1536–1548.

Chonka, A., & Abawajy, J. (2012, September). Detecting and mitigating HX-DoS attacks against cloud web services. In *2012 15th International Conference on Network-Based Information Systems* (pp. 429-434). IEEE. 10.1109/NBiS.2012.146

Costa-Requena, J., Santos, J. L., Guasch, V. F., Ahokas, K., Premsankar, G., Luukkainen, S., ... de Oca, E. M. (2015, June). SDN and NFV integration in generalized mobile network architecture. In *2015 European conference on networks and communications (EuCNC)* (pp. 154-158). IEEE. 10.1109/EuCNC.2015.7194059

Dierks, T., & Rescorla, E. (2008). *The transport layer security (TLS) protocol version 1.2*. Academic Press.

Elijah, O., Leow, C. Y., Rahman, T. A., Nunoo, S., & Iliya, S. Z. (2015). A comprehensive survey of pilot contamination in massive MIMO—5G system. *IEEE Communications Surveys and Tutorials, 18*(2), 905–923. doi:10.1109/COMST.2015.2504379

Fonseca, P., Bennesby, R., Mota, E., & Passito, A. (2012, April). *A replication component for resilient OpenFlow-based networking. In 2012 IEEE Network operations and management symposium*. IEEE.

Gember, A., Dragga, C., & Akella, A. (2012, October). ECOS: Leveraging software-defined networks to support mobile application offloading. In *2012 ACM/IEEE Symposium on Architectures for Networking and Communications Systems (ANCS)* (pp. 199-210). IEEE. 10.1145/2396556.2396598

Goyal, S., Liu, P., Panwar, S. S., Difazio, R. A., Yang, R., & Bala, E. (2015). Full duplex cellular systems: Will doubling interference prevent doubling capacity? *IEEE Communications Magazine, 53*(5), 121–127. doi:10.1109/MCOM.2015.7105650

Jaeger, B. (2015, August). *Security orchestrator: Introducing a security orchestrator in the context of the etsi nfv reference architecture. In 2015 IEEE Trustcom/BigDataSE/ISPA* (Vol. 1). IEEE.

Jafarian, J. H., Al-Shaer, E., & Duan, Q. (2012, August). Openflow random host mutation: transparent moving target defense using software defined networking. In *Proceedings of the first workshop on Hot topics in software defined networks* (pp. 127-132). 10.1145/2342441.2342467

Khan, A. N., Kiah, M. M., Khan, S. U., & Madani, S. A. (2013). Towards secure mobile cloud computing: A survey. *Future Generation Computer Systems, 29*(5), 1278–1299. doi:10.1016/j.future.2012.08.003

Kim, H., Jung, I., Park, Y., Chung, W., Choi, S., & Hong, D. (2018). Time spread-windowed OFDM for spectral efficiency improvement. *IEEE Wireless Communications Letters, 7*(5), 696–699. doi:10.1109/LWC.2018.2812150

Kulkarni, P., Khanai, R., & Bindagi, G. (2016, March). Security frameworks for mobile cloud computing: A survey. In 2016 international conference on electrical, electronics, and optimization techniques (ICEEOT) (pp. 2507-2511). IEEE. doi:10.1109/ICEEOT.2016.7755144

La Polla, M., Martinelli, F., & Sgandurra, D. (2012). A survey on security for mobile devices. *IEEE Communications Surveys and Tutorials, 15*(1), 446–471. doi:10.1109/SURV.2012.013012.00028

Lauer, H., & Kuntze, N. (2016, July). Hypervisor-based attestation of virtual environments. In *2016 Intl IEEE Conferences on Ubiquitous Intelligence & Computing, Advanced and Trusted Computing, Scalable Computing and Communications, Cloud and Big Data Computing, Internet of People, and Smart World Congress (UIC/ATC/ScalCom/CBDCom/IoP/SmartWorld)* (pp. 333-340). IEEE. 10.1109/UIC-ATC-ScalCom CBDCom-IoP-SmartWorld.2016.0067

Lu, L., Li, G. Y., Swindlehurst, A. L., Ashikhmin, A., & Zhang, R. (2014). An overview of massive MIMO: Benefits and challenges. *IEEE Journal of Selected Topics in Signal Processing, 8*(5), 742–758. doi:10.1109/JSTSP.2014.2317671

Namal, S., Ahmad, I., Gurtov, A., & Ylianttila, M. (2013, November). *Enabling secure mobility with OpenFlow. In 2013 IEEE SDN for Future Networks and Services (SDN4FNS)*. IEEE.

Namal, S., Ahmad, I., Saud, S., Jokinen, M., & Gurtov, A. (2016). Implementation of OpenFlow based cognitive radio network architecture: SDN&R. *Wireless Networks, 22*(2), 663–677. doi:10.100711276-015-0973-5

Neisse, R., Costa, P. D., Wegdam, M., & van Sinderen, M. (2008, June). An information model and architecture for context-aware management domains. In *2008 IEEE Workshop on Policies for Distributed Systems and Networks* (pp. 162-169). IEEE. 10.1109/POLICY.2008.31

Norrman, K., Näslund, M., & Dubrova, E. (2016, June). Protecting IMSI and user privacy in 5G networks. In *Proceedings of the 9th EAI international conference on mobile multimedia communications* (pp. 159-166). 10.4108/eai.18-6-2016.2264114

Park, Y., & Park, T. (2007a). A survey of security threats on 4G networks. *Proceedings of the IEEE Globecom Workshops*, 1–6. 10.1109/GLOCOMW.2007.4437813

Park, Y., & Park, T. (2007b, November). *A survey of security threats on 4G networks. In 2007 IEEE Globecom workshops*. IEEE.

Santos, M. A., De Oliveira, B. T., Margi, C. B., Nunes, B. A., Turletti, T., & Obraczka, K. (2013, October). Software-defined networking based capacity sharing in hybrid networks. In *2013 21st IEEE international conference on network protocols (ICNP)* (pp. 1-6). IEEE. 10.1109/ICNP.2013.6733664

Security, S. D. N. (2013). *Considerations in the Data Center*. Open Networking Foundation. Available: https://www.opennetworking.org/sdn-resources/sdn-library

Strassner, J. (2004). Chapter 4-policy operation in a PBNM system. Policy-Based Network Management.

Sucasas, V., Mantas, G., & Rodriguez, J. (2016). Security challenges for cloud radio access networks. *Backhauling/Fronthauling for Future Wireless Systems*, 195-211.

Suo, H., Liu, Z., Wan, J., & Zhou, K. (2013, July). Security and privacy in mobile cloud computing. In *2013 9th International Wireless Communications and Mobile Computing Conference (IWCMC)* (pp. 655-659). IEEE. 10.1109/IWCMC.2013.6583635

Töyssy, S., & Helenius, M. (2006). About malicious software in smartphones. *Journal in Computer Virology*, 2(2), 109–119. doi:10.100711416-006-0022-0

Twidle, K., Dulay, N., Lupu, E., & Sloman, M. (2009, April). Ponder2: A policy system for autonomous pervasive environments. In *2009 Fifth International Conference on Autonomic and Autonomous Systems* (pp. 330-335). IEEE. 10.1109/ICAS.2009.42

Vikas, S. S., Pawan, K., Gurudatt, A. K., & Shyam, G. (2014, February). Mobile cloud computing: Security threats. In 2014 international conference on electronics and communication systems (ICECS) (pp. 1-4). IEEE.

Vook, F. W., Ghosh, A., & Thomas, T. A. (2014, June). MIMO and beamforming solutions for 5G technology. In *2014 IEEE MTT-S International Microwave Symposium (IMS2014)* (pp. 1-4). IEEE. 10.1109/MWSYM.2014.6848613

YuHunag, C., MinChi, T., YaoTing, C., YuChieh, C., & YanRen, C. (2010, November). A novel design for future on-demand service and security. In *2010 IEEE 12th International Conference on Communication Technology* (pp. 385-388). IEEE.

Zhao, C., Huang, L., Zhao, Y., & Du, X. (2017). Secure machine-type communications toward LTE heterogeneous networks. *IEEE Wireless Communications*, 24(1), 82–87. doi:10.1109/MWC.2017.1600141WC

Zhao, H., Lobo, J., & Bellovin, S. M. (2008, June). An algebra for integration and analysis of ponder2 policies. In *2008 IEEE Workshop on Policies for Distributed Systems and Networks* (pp. 74-77). IEEE. 10.1109/POLICY.2008.42

Zheng, Y., He, D., Yu, W., & Tang, X. (2005, December). Trusted computing-based security architecture for 4G mobile networks. In *Sixth International Conference on Parallel and Distributed Computing Applications and Technologies (PDCAT'05)* (pp. 251-255). IEEE. 10.1109/PDCAT.2005.243

Zhou, J., Shen, Q., & Xu, Y. (2012, May). Research and improvement of Ponder2 policy language. In *2012 IEEE International Conference on Computer Science and Automation Engineering (CSAE)* (Vol. 2, pp. 455-458). IEEE. 10.1109/CSAE.2012.6272813

Chapter 3
Machine Learning Algorithms for 6G Wireless Networks:
A Survey

Anita Patil
S. G. Balekundri Institute of Technology, India

Sridhar Iyer
iD https://orcid.org/0000-0002-8466-3316
KLE Dr. M. S. Sheshgiri College of Engineering and Technology, India

Rahul J. Pandya
Indian Institute of Technology, Dharwad, India

ABSTRACT

Over the past decade, in view of minimizing network expenditures, optimizing network performance, and building new revenue streams, wireless technology has been integrated with artificial intelligence/machine learning (AI/ML). Further, there occurs dramatic minimization of power consumption and improvement in system performance when traditional algorithms are replaced with deep learning-based AI techniques. Implementation of ML algorithms enables wireless networks to advance in terms of offering high automation levels from distributed AI/ML architectures applicable at network edge and implementing application-based traffic steering across access networks. This has enabled dynamic network slicing for addressing different scenarios with varying quality of service requirements and has provided ubiquitous connectivity across various 6G communication platforms. Keeping a view of the aforementioned, in this chapter, the authors present a survey of various ML techniques that are applicable to 6G wireless networks. They also list open problems of research that require timely solutions.

DOI: 10.4018/978-1-6684-3921-0.ch003

INTRODUCTION

Overview of machine learning in wireless communication networks

With the exponential increase in the bandwidth demand and data traffic, there is an immediate requirement to serve this traffic through high-speed wireless communication networks. In turn, this requisites efficient software enabled intelligent algorithms, advanced physical layer solutions, and spectral bands at a higher frequency to fulfil the requirements of the next-generation users. The wireless communication research community has recently shown that the Tera-Hertz (THz) band is one of the promising bands to enable ultra-broadband wireless communication and minimize the spectrum scarcity issues (Zhao et. al., 2021).

The current wireless systems rely heavily on mathematical models; however, such models do not define the system structure accurately. Hence, the use of Machine learning (ML) techniques for wireless communication has gained momentum as these methods enable the attainment of the quality of service functionalities with advanced solutions (Ali et. al., 2020). Moreover, ML techniques provide the replacement of heuristic or Brute Force Algorithms for optimizing localized tasks and can also present adequate solutions that the existing mathematical model are unable to obtain. Currently, the ML algorithms are being deployed and trained statically at different management layers, core, radio base stations, and mobile devices. The dynamic deployment is envisioned to yield enhanced performance and utilization.

In general, the ML algorithms help in tasks such as, classification, regression, the interaction of an intelligent agent with the wireless environment (Syed et. al. 2019). In such operations, ML algorithms work in three different versions viz., supervised learning, unsupervised learning, and reinforcement learning. Few ML models such as, non-parametric Bayesian methods (Gaussian approach), are promising, especially in handling small, incrementally growing data sets; however, they have increased complexity compared to the parametric methods. Further, the Kernel Hilbert Space-based solutions have shown encouraging results in generating improved data rate, which is 10-100 times higher in comparison to the ones shown in the 5G wireless networks, simultaneously being computationally simple and scalable with lower approximation error. Federated Learning (FL) is an alternate distributed ML algorithm which enables mobile devices to collaboratively learn a shared ML model without data exchange among mobile devices (Marmol et. al., 2021). It is being analysed further to be considered as a next-generation solution for orientation, intrusion detection, mobility, and extreme event prediction. Reinforced Learning algorithms help in coding scheme selection, modulation, beam forming, and power control. In addition, physical layer optimization also exploits ML for multi-input and multi-output downlink beam forming. Implementing all the aforementioned ML algorithms at the end-user devices, needs the consideration of key parameters such as, cost, size, and power. Additional considerations in the simulation and the prototyping of ML at the end-user devices are to optimize the physical realization of the design and finding the inputs to the model (Dalal & Kushal, 2019).

The focus of this chapter is to bring out the importance of AI and ML in 6G wireless communication. ML is a component of AI although it endeavors to solve the problems based on historical or previous examples. Unlike AI applications, ML involves learning of hidden patterns within the data (data mining) and subsequently using the patterns to classify or predict an event related to the problem. Simply put, intelligent machines depend on the knowledge to sustain their functionalities and ML offers the same. In essence, ML algorithms are embedded into machines and data streams provided so that knowledge and information are extracted and fed into the system for faster and efficient management of processes (Ali et. al., 2020).

Demand for radio spectrum is increasing as the data traffic is increasing, and hence, massive connections with high quality of service have to be provided. Recent advances in ML have shown that ML will play a major role in solving multiple issues in wireless communication networks. To mention few, ML will provide ease in all sort of applications which were not enabled in the earlier generations such as, Augmented Reality (AR), Virtual Reality (VR), holographic telepresence, eHealth, wellness applications, Massive Robotics, Pervasive connectivity in smart Environment, etc. It is envisioned that ML will enable real time analysis and zero-touch operation, and will provide control in in the 6G networks (Zhao et. al., 2020). Mobile devices can assist and report to the network regarding the ML actions and predictions to aid efficient resource management. In order to manage the connection density, dynamic spectrum management has been proposed in the literature. The key enabling techniques for dynamic spectrum are i) Cognitive Radio ii) Symbiotic Radio, and iii) Blockchain Technology (Hong et. al., 2014; Hewa et. al. 2020).

The scarcity of available spectrum and underutilization of the allocated spectrum necessitates efficient techniques to manage the spectrum dynamically. In dynamic spectrum management, the concept of primary and secondary users exists wherein; secondary users do not have the authority to access the spectrum; however, they can access it whenever the primary spectrum is idle, and it can even be shared with the protection of primary users' service. This process enables the secondary users to transmit their data without the licence of spectrum.

In order to achieve dynamic spectrum allocation many algorithms are proposed in the literature. These algorithms not only address the issues of spectrum allocation but also issues such as, data security, optimization, power-efficiency, cost-efficiency, etc. Following are the ML algorithms which help in addressing all the aforementioned issues:

1. Supervised Learning
2. Unsupervised learning
3. Reinforced Learning
4. Federated Learning
5. Kernel Hilbert Space
6. Block Chain Technology
7. Cognitive Radio
8. Symbiotic Radio.
9. THz Technology
10. Free Duplex
11. Index Modulation

SUPERVISED LEARNING AND UNSUPERVISED LEARNING ALGORITHMS

The ML algorithms can be mainly classified as supervised or unsupervised. These two classes mainly differ in the labels of training data sets. In supervised ML output and input attributes are predetermined (Amanpreet et. al. 2016). The algorithms perform prediction and classification of the predetermined attributes, and their accuracy. For instance, if the input variable is X and output variable is Y then, the mapping function using by supervised learning will be $Y=F(X)$. The learning process stops when the algorithm achieves an acceptable level of performance. As detailed by (Chowdhury et. al., 2020), first analytical tasks are performed by supervised algorithms using the training data, and subsequently contingent functions are constructed for mapping variations of the attribute. These algorithms need pre-specifications of maximum settings to obtain the desired outcome and performance. With this approach in ML, it has been observed that the training subset of approximately 66% is rationale, and without demanding more computational time, outcome will be achieved. Further, the supervised learning algorithms can be further classified into the classification and regression algorithms:

- **Classification**: If output variable is a category such as, "red" or "blue" or "disease" and "no disease" the classification problem can be used (Amanpreet et. al. 2016).
- **Regression**: If the output value is a real value such as, "rupees" then, the regression problem can be used. Few popular examples of supervised ML algorithms are:
  - Linear regression for regression problems.
  - Random forest for classification and regression problems.

- ○ Support vector machines (SVM) for classification problems.
- ○ For resource allocation, and coding scheme selection in wireless communication, supervised learning algorithms are utilized.

Semi-Supervised Machine Learning

In these algorithms, large amount of input data X and small amount of output data Y are labelled. These problems lie in between the supervised and the unsupervised learning (Sheena & Sachin, 2019; Yogesh et. al., 2020).

Unsupervised Learning

In contrast to the supervised learning, unsupervised data learning comprises of pattern recognition without having a target attribute. All the variables used in the analysis are used as inputs, and the techniques are suitable for clustering and association mining techniques. According to (Amruthnath at. al., 2018), unsupervised learning algorithms are used to create labels in the data that are subsequently used to implement the supervised learning tasks. Clustering algorithms identify inherent groupings within the unlabeled data, and subsequently assign label to each data value. Further, unsupervised association mining algorithms identify rules that accurately represent the relationships between the attributes. Overall, the aim of unsupervised learning is to model the underlying structure or distribution in the data so as to learn more about the data. These algorithms do provide correct answers and there is no teacher; hence, they are called as 'Unsupervised Algorithms'. These algorithms can also be grouped into clustering and association algorithms (Zhao et. al., 2021).

- **Clustering**: To discover the inherent groupings in the data. An example is based on purchasing behaviour grouping of customers.
- **Association**: To discover rules that describe large portions of the given data such as, people who buy X will also buy Y.

Few additional examples of the unsupervised learning Algorithms are:

- K-means for clustering problems.
- A priori algorithm for association rule learning problems.
- The commonly used techniques are: clustering, auto encoders (Kien et. al., 2020), deep belief nets, generative adversarial networks, and the expectation–maximization algorithm. It is also used in the physical layer for optimal modulation, channel-aware feature extraction, anomaly detection, localization etc.

REINFORCEMENT ALGORITHM

Reinforcement learning (RL) is one of the basic ML paradigms along with supervised and unsupervised learning. It is concerned with the manner in which intelligent devices are required to make decisions or take actions in an environment in order to maximize the notion of cumulative reward. It differs from supervised learning as it does not need labelled input/output pairs to be presented, and does not need sub-optimal actions to be explicitly corrected. Instead, the focus is on finding a balance between exploration and exploitation. Markov Decision Process (MDP) is typically used to state the environment, as reinforcement learning algorithms make use of dynamic programming techniques. Reinforcement learning algorithms do not assume knowledge of an exact mathematical model of the MDP; however, they target large MDPs where exact mathematical models become infeasible (Vasileios. Et. al., 2021).

RL finds an application in many disciplines such as, game theory, control theory, simulation based optimization, multi-agent systems, swarm intelligence and wireless communication. For example, in the operational research and control literature, reinforcement learning is referred as *approximate dynamic programming,* or *neuro-dynamic programming.* The theory of optimal control includes reinforcement learning problems which is concerned with the characterization and existence of optimal solutions and algorithms for their exact computation. The problems are less concerned with approximation or learning, specifically in the absence of a mathematical model of the environment. Function approximation and use of samples are the powerful elements of reinforcement learning use to deal with large environments and optimization of performance. Reinforcement learning problem consists of an agent to interact with an environment and learns how to take actions. At each step of the learning process, state of the environment is observed by the agent which then takes action from the available set of actions, receives a numeric reward, and moves to the next state. Hence, the aim of the agent is to maximize long term cumulative reward. Problem such as resource allocation in wireless domain can be formulated as a reinforcement learning problem wherein, neural networks can be used as function approximators to learn values of each state or the rewards that are generated by the environment. Many problems in wireless networks such as, power control, beamforming, and modulation and coding scheme selection are solved by various deep reinforcement learning architectures. However, as a drawback, reliance on training is a major limitation of reinforcement learning. In this regard, recently, there has been advances towards reducing this reliance, specifically when dealing with extreme network situations. In particular, the concept of experienced deep reinforcement learning was proposed in which reinforcement learning is trained using Generative Adversarial Networks (GAN) that generate synthetic data to complement a limited, existing real dataset (Iqbal, 2021).

Two elements make RL powerful viz., the use of samples to optimize the performance and the use of function approximation to deal with large environments. RL can be used in following situations:

1. Only a simulation model of the environment is known i.e., subject of simulation optimization.
2. A model of the environment is known and analytic solution is not available
3. In 6G, RL can be used for resource allocation in cognitive network, beam forming, coding scheme selection, power control, channel modulation, etc.

RL Algorithms can also be categorised as described in the following sections.

Associative Reinforcement Learning

The learning system interacts in a closed loop with its environment and combines the facets of stochastic learning automata tasks.

Deep Reinforcement Learning

RL is extended in this approach by using a deep neural network without explicitly designing the state space. Learning ATARI games by Google Deep Mind has created increased attention to Deep Reinforcement Learning (DRL). It influences Markov decision models for selecting next 'action' based on the state transition models. In DRL, instead of mapping every solution, states are approximated or estimated by a neural network. In 6G wireless communications, efficient solutions for radio resource allocation are required and this is challenging as 6G wireless network will aim to serve wider variety of users in future radio resources which will be in extreme scarcity. As a solution, DRL can be implemented to obtain efficient solutions for the radio resource problems (Zhao et. al., 2020).

Inverse Reinforcement Learning

Purpose of inverse reinforcement learning is to imitate observed behaviour which is often optimal solution or close to optimal. Inverse reinforcement learning approach will have no reward function but a reward function will be inferred as an observed behaviour from expert.

Safe Reinforcement Learning

Safe RL is the process of learning policies that maximize the expectation of the return in problems wherein, it is important to ensure reasonable system performance, safety constraints during learning and deployment processes.

FEDERATED LEARNING

Federated learning (FL) is helpful is cases when it is difficult to assign the ML models to each mobile device and data center because in traditional centralized ML algorithms, it is required for each mobile device to transmit its collected data to the data center for training purpose. It is impractical for wireless mobile devices to transmit their local data for training ML models due to privacy issues. To overcome this traditional ML problem, FL, which is a distributed ML algorithm that enables devices to learn a shared ML model without data exchange among mobile devices, is implemented (Li et. al., 2021). In this approach, each mobile device and the data center will have their own ML models, and these ML models are referred to as 'local Model' and 'Global Model' for mobile device and data center, respectively.

Further, the training model of ML in FL approach is as follows:

1. Each mobile device uses the collected data to train the local FL model and sends the trained local FL model to the data center.
2. The data center integrates the local FL models to generate the global FL model and broadcasts it back to all mobile devices.

As it can be inferred from the FL process, the mobile devices are required to transmit the training parameters over the wireless links. The limited wireless bandwidth, imperfect transmission and dynamic wireless channels will affect the FL performance in a significant manner. However, many studies have shown that implementation of FL process by optimizing the wireless network. The main objective of FL is to protect the data owner's privacy. It aims to train a ML model with training data kept distributed at the clients in order to protect the data owners' privacy. The working idea illustration includes devices of the users are used to train local models and then the trained local models are sent to the base station for aggregation. In this process, the privacy of the user data is well preserved as the data are still maintained in the devices. In 6G networks, the architecture will be distributed wherein, the FL technology of AI from centralized cloud based model to decentralized devices will be necessitated. The AI computing tasks can be distributed to multiple decentralized edge nodes from a central node, and hence, FL is one of the vital ML methods to enable the deployment of accurately generalized models across multiple devices (Du et. al., 2020).

KERNEL HILBERT SPACE

High interference, which is a result of massive connectivity, will be a major performance bottleneck in 6G. Massive connectivity will involve serving extensive range of devices of various manufacturing qualities and this will lead to introduction of impairments which develop due to diverse objects introduced by non-ideal hardware. The major objects may be I/Q imbalance, non-linear characteristics, high-mobility especially in the industries where fixed solution may not be applicable, etc. To accomplish the promise of improvement in the data rate of 10-100 times in 6G as compared to the scenarios of 5G, the Reproducing kernel Hilbert space (RKHS) based solutions are predominantly useful owing to their computational simplicity (Salh et. al., 2021). RKHS methods have significantly lower approximation error and are scalable. 6G will potentially encounter the high interference non-Gaussian environments; however, RKHS based solutions will provide efficiency with lower approximation error. RKHS based approaches have recently appeared as a solution for many impairments in the framework of numerous applications in the next-generation wireless communication systems. Hence, the major issues in 6G such as, detection, tracking and localization will be addressed by various RKHS based solutions. In recent years, technological advances have ensured that DL method offers tremendous solutions, and it is largely used in wireless communications problems. Simultaneously, further improvement in the performance of RKHS based approaches could be achieved by the feature extraction methods using Monte-Carlo sampling. The extracted features can be utilized as input to the DL based slants to boost the act of models used in 6G. As compared to the classical DL algorithms, RKHS based DL algorithms ensure enhanced performance as there is an intrinsic regularization and support with strong analytical framework (Guo, 2021).

COGNITIVE RADIO

The most widely accepted key technology to empower dynamic spectrum allocation is cognitive radio. Based on the interaction with surrounding environment, and also based on the awareness of its internal states such as, hardware and software architectures, user needs, spectrum use policy, cognitive radio can autonomously and dynamically adapt the system transmission strategies, bandwidth, transmit power, antenna beam, carrier frequency and modulation scheme. The transmission strategies are adjusted by computer software in the Software Defined Radio. Cognitive radio is capable to observe, analyze the observed environment through sensing, and process information through learning. It decides the transmission strategy through reasoning. Recent advances in research have shown that cognitive radio technology can be explored more for its inherent potential towards AI, and also its capacity to facilitate Dynamic Spectrum Allocation (DSA) very efficiently (Hong et. al., 2014).

Although most of the existing cognitive radio researchers till date have been focusing on the exploration and realization of cognitive capability to facilitate the DSA, recent research has been conducted to explore additional potential which is inherent in the cognitive radio technology through AI. A typical cognitive cycle for cognitive radio comprises of a Secondary User having cognitive capability which is essential to consistently and periodically observe the environment, and it obtains the information regarding interference temperature, spectrum holes, etc. It also determines the best operational features to optimize its own performance subjected to protecting the primary users, and according to these operational parameters, system configurations are conducted. The traffic statistics and channel fading statistics of the radio environment can be analyzed by the information time. Due to this, the cognitive radio equipment is able to learn and perform better in future Dynamic Spectrum Adaptation (DSA). However, implementing DSA with cognitive radio involves efforts from various research communities such as, communications, signal processing, computer networking, information theory and ML. Lastly, combining DSA with cognitive radio and its realization also fundamentally hinge on the inclination of regulators to open the spectrum for unlicensed access. Providentially, over the past decades, it has been observed that universal determinations from regulatory bodies are eradicating regulatory barriers to expedite DSA (Mollah et. al., 2019).

BLOCKCHAIN TECHNOLOGY

In addition to the cryptocurrency, with its prominent features, blockchain has many uses including, smart contracts, financial services and IoT. Besides, blockchain can fetch new openings to improve the competence, and to lessen charge in the dynamic spectrum management. Also, blockchain is utilized to manage and share the spectrum resources, due to which central authority may get eliminated. Blockchain is fundamentally a distributed and open archive, in which transactions are firmly recognized in blocks. In the existing block, an exclusive indicator determined by transactions in the preceding block is recorded. Tampering with any transaction deposited in a previous block can be detected efficiently. The transactions commenced by one node are broadcast to other nodes, and a consensus algorithm is opted to conclude which node is approved to authorize the new block by affixing it to the blockchain. With the distributed authentication and record mechanism, blockchain will be robust against single point of failures, verifiable and transparent. Depending on the level decentralization, blockchain can be classified into consortium blockchain, public blockchain and private blockchain. A public blockchain can be tested and retrieved by all nodes in the network; whereas, a private blockchain or a consortium blockchain maintained by the permissioned nodes only. A smart contract, reinforced by the blockchain technology, is a self-executable connection with its sections being converted to programming writings and deposited in a transaction. Blockchain has been explored to provide support to various applications of IoT. As a decentralization entity, blockchain can be useful in integrating the various IoT devices and strongly stock the massive data formed by them. Moreover, blockchain is also used to accomplish the mobile edge computing resources to preserve the IoT devices with limited computation capacity. Blockchain provides security to dynamic spectrum access enabled by the spectrum auctions. In opportunistic spectrum access provided by sensors, block chain is used to intermediate the spectrum sensing service (Abdualgalil et. al., 2020). Specifically, blockchain can be used to:

1. **Establish a self-organized Spectrum Market:** In the management of spectrum resources, a self-organized spectrum market is essential with enhanced efficiency, with the reduced cost when compared to depending on a centralized authority. Dynamic spectrum management, transactions and their verification and smart contract combination will be used to block chain based self-organized spectrum market. The blockchain security assures that users can make the transactions without trust in each other. Thus, high bar to achieve spectrum resource accessibility can be reduced.

2. **Create a secured database**: The information such as, outcomes of spectrum auctions and access records, historical sensing results can also be stored in the blockchain.

Overall, block chain technologies have been assumed to convey new opportunities to the dynamic spectrum management, to ensure progress in the avenues of security, autonomy, decentralization and administration cost reduction. Moreover, investigation on challenges such as, deployment, energy consumption, and scheme of blockchain network over the conventional cognitive radio network should also be carried out.

SYMBIOTIC RADIO

Symbiotic radio (SR) is an innovative technique that proposes the rewards and overcomes shortcomings of ambient backscattering communications (AmBC) and cognitive radio (CR). In building 6G as a spectrum and energy efficient communication system, SR is considered as one of the promising approach. In addition, Intelligent Reflecting Surfaces (IRS tool) can further play a role in improving the concert of the transmission by enhancing the backscattering link signal (Mingzhe et. al., 2021). Additional technologies, related to spectrum, in 6G are as follows:

Terahertz

Terahertz (THz) is measured as one of the vital technologies for the 6G wireless communications. 5G is described under 100 GHz as the millimeter-wave bands, whereas, 100 GHz-3 THz is considered as the THz band in 6G. The beyond 90 GHz band is purely used for scientific service which has not been fully discovered. Therefore, it is proposed to support the increased wireless network capacity. THz also empowers the ultra-low latency communication and ultra-high bandwidth paradigms, which caters the desires of several evolving applications such as, IoT and autonomous driving. THz is especially adequate for high bit-rate short-range communications since, path loss increases with increase in frequency which makes it challenging for long-range communications.

Free Duplex

6G will eradicate the difference between Frequency Division Duplex and Time Division Duplex, and the frequency sharing will be grounded on the necessities, which is known as free duplex. Hence, the spectrum resource allocation in 6G will be more effective and efficient. With the Free duplex technology, transmission rate and throughput in 6G can be increased and transmission latency will be reduced (Zhao et. al., 2021).

Index Modulation

Index modulation (IM) can progress the transmission rate as it carries the source information bits through the classical amplitude-phase modulation (APM) signals. Therefore, it can potentially be used in 6G. Studies suggest that information bits can be communicated through the index of the antennas in MIMO systems. This technique is known as space shift keying (SSK), and if SSK is combined with classical linear modulation, it is called as amplitude-phase modulation (APM). For example, space modulation (SM) is proposed based on the same idea as in SSK. In SM technology, the base information bits are distributed into two parts: the index of the transmit antennas and the other parts for the APM. Consequently, SM can meaningfully increase the transmission rate by sending the extra information bits through the traditional APM transceiver's antenna index. On one side of the antennas, further resource units can also be indexed to transmit the additional information bits. These resource units include, channel state, sub-carriers, and time slots. Index Modulation, along with Orthogonal frequency division multiplexing access (OFDMA), will be the key technology to suggestively upsurge the quantity output for supporting more users to access the 6G network (Zhao et. al., 2020).

Overall, 6G is expected to be a composite network where a huge selection of smart devices is connected to the system and are mandatory to interconnect with others anytime, and the life period of the battery-charging components is also vital to fulfill the restraints of ultra-low power consumption. To extend the life duration of numerous devices in the network, simultaneous wireless information and power transfer (SWIPT) technology is anticipated. SWIPT empowers sensors to be charged using wireless power transfer; thereafter, battery-free devices can be maintained in 6G, dropping the network's power consumption considerably. Consequently, performance on the sum rate, throughput, and outage probability for non-orthogonal multiple access (NOMA) networks with SWIPT are derived.

APPLICATIONS OF 6G

In the case of 6G networks, a varied series of AI applications will develop into 'connected intelligence (Shrestha et. al., 2019), later smoothing every facet of our day-to-day life. For instance:

1. Innovative AI methods can be hired in autonomy to save manpower or for network management. Overall, 6G invigorates smart healthcare by providing high-precision medical treatment, real-time health monitoring, and reliable privacy protection.
2. With the arrival of 6G, Industry 4.0 will be completely comprehended as smart manufacturing and it will achieve high precision manufacturing.
3. Intelligent robots coupled by pervasive 6G network empower manufacturing structures to carry out multifaceted and hazardous tasks devoid of risking people's life.
4. The smart home that furnishes with smart IoT devices will deliver a contented living atmosphere to people, and 6G permits the smart home to confirm the occupants' safety.
5. In terms of traffic and transportation, the sophisticated sensing and planning algorithms can be deployed for traffic optimization.
6. Other applications such as, smart grid and unmanned aerial vehicle will also be enhanced with the aid of 6G.

Explainable Artificial Intelligence

An enormous number of applications such as, remote surgery and autonomous driving exist in 6G eon. As these applications are attentively connected to humans' life, a small error may invite miserable accidents. Hence, it is very essential to mark an AI explainable for building faith between individuals and machines. Currently, most AI tactics in MAC and PYH layers of 5G wireless networks are incomprehensible. AI applications such as, remote surgery and autonomous driving will be an extensive part of 6G, which needs explaining ability to qualify trust. AI decisions should be understood by human experts, and must be explainable so as to be considered as reliable. Prevailing approaches, including didactic statements, visualization with case studies, hypothesis testing can improve the explaining ability of DL (Guo, 2019; Abhishek & Neha, 2021).

CONCLUSION

Open Research Avenues

Finally, we list the various open problems of research in regard to ML in 6G wireless networks

- Implementation of automation requires higher data rates, flexibility, link preservation, scheduling, on demand beam forming-deep neural network with reinforcement learning algorithms need to be analysed for usage and deployment constraints have to be studied. In such case, solution to a challenge 'how to use deep reinforcement learning for the automation of 6G wireless network' will be an observable one.
- Ultra-reliable low latency and interference management are the major requirements in open data access. To support the emerging machine critical applications, the transmission time delay of E-to-E (Equipment to Equipment) is expected to be less than 1ms. Satisfying low latency and high reliability needs with open data access in business oriented mobile network is an open problem in 6G.
- In Application and Platform dependent models a design of device is necessary which can sense the environment and its history so that resources can be allocated dynamically and spectrum selection should also happen dynamically to avoid the congestion. Hence the transfer of models to highly resource constrained platforms and designing a method to dynamically select application and platform based models is a research challenge.

- Implementation of ML algorithms for enhancing, automating and managing the 6G network performance is another research challenge.

REFERENCES

Abdualgalil, B., & Abraham, S. (2020). Applications of Machine Learning Algorithms and Performance Comparison: A Review. *IEEE 2020 International Conference on Emerging Trends in Information Technology and Engineering (ic-ETITE)*, 1–6. 10.1109/ic-ETITE47903.2020.490

Abhishek, D., & Neha, G. (2021). A Systematic Review of Techniques, Tools and Applications of Machine Learning. *Proceedings of the Third International Conference on Intelligent Communication Technologies and Virtual Mobile Networks (ICICV 2021)*. 10.1109/ICICV50876.2021.9388637

Ali, S., Saad, W., & Rajatheya. (2020). *6G White Paper on Machine Learning in Wireless Communication Networks*. arXiv:2004.13875v1.

Amanpreet, S., Narina, T., & Aksha, S. (2016). A review of supervised machine learning algorithms. *3rd International Conference on Computing for Sustainable Global Development (INDIACom)*.

Amruthnath, N., & Gupta, T. (2018). A research study on unsupervised machine learning algorithms for early fault detection in predictive maintenance. *5th International Conference on Industrial Engineering and Applications (ICIEA)*. 10.1109/IEA.2018.8387124

Chowdhury, S., & Schoen, M. P. (2020). *Classification using Supervised Machine Learning Techniques. Intermountain Engineering Technology and Computing*. doi:10.1109/ietc47856.2020.924921

Dalal, R., & Kushal, R. (2018). Review on Application of Machine Learning Algorithm for Data Science. *IEEE 3rd International Conference on Inventive Computation Technologies (ICICT)*, 270-273. 10.1109/ICICT43934.2018.9034256

Du, J., & Jiang, C. (2020). Machine Learning for 6G Wireless Networks: Carry-Forward-Enhanced Bandwidth, Massive Access, and Ultrareliable/Low Latency. *IEEE Vehicular Technology Magazine*. Advance online publication. doi:10.1109/MVT.2020.3019650

Guo, W. (2019). *Explainable artificial intelligence (XAI) for 6G: Improving trust between human and machine*. arXiv preprint arXiv:1911.04542.

Hewa, T., Gurkan, G., Kalla, A., & Ylianttila. (2020). The Role of Blockchain in 6G: Challenges, Opportunities and Research Directions. *2nd 6G wireless Summit (6G SUMMIT)*. . doi:10.1109/6GSUMMIT49458.2020.9083784

Hong, X., Jing, W., & Jianghong, S. (2014). Cognitive radio in 5G: A perspective on energy-spectral efficiency trade–off. *IEEE Communications Magazine, 52*(7), 46–53. Advance online publication. doi:10.1109/MCOM.2014.6852082

Iqbal, H. S. (2021). Machine Learning: Algorithms, Real World Applications and Research Directions. *SN Computer Science, 2*(160). doi:10.1007/s42979-021-00592-x

Kien, T., Long Ton, T., & Nguyen, G. M. T. (2020). Plant Leaf Disease Identification by Deep Convolutional Autoencoder as a Feature Extraction Approach. *Proceedings of The IEEE 17th International Conference on Electrical Engineering / Electronics, Computer, Telecommunications and Information Technology (ECTI-CON)*. Available on https://ieeexplore.ieee.org/document/9158218/

Li, T., Sahu, A. K., Talwalkar, A., & Smith, V. (2020). Federated Learning: Challenges, Methods, and Future Directions. *IEEE Signal Processing Magazine, 37*(3), 50–60. doi:10.1109/MSP.2020.2975749

Marmol, C., Pablo, F. S., Aurora, G. V., Jose, L., Hernandez, R., Jorge, B., Gianmarco, B., & Antonio, S. (2021). *Evaluating Federated Learning for Intrusion Detection in Internet of Things: Review and Challenges Enrique*. arXiv:2108.00974v1.

Mingzhe, C., Deniz, G., Kaibin, H., Walid, S., & Mehdi, B. (2021). Distributed Learning in Wireless Networks: Recent Progress and Future Challenges. *IEEE Journal on Selected Areas in Communications, 39*(12), 3579–3605. doi:10.1109/JSAC.2021.3118346

Mollah, M. B., Zeadally, S., & Azad, M. A. K. (2019). *Emerging wireless technologies for Internet of Things applications: Opportunities and challenges. In Encyclopaedia of Wireless Networks*. Springer International Publishing Cham.

Salh, A., Audah, L., & Shah, N. S. M. (2020). A Survey on Deep Learning for Ultra-Reliable and Low-Latency Communications Challenges on 6G Wireless Systems. *Proceedings of Future of Information and Communication Conference (FICC) 2021*. arXiv: 2004.08549v3.

Sheena, A., & Sachin, A. (2017). Machine learning and its applications: A review. *International Conference on Big Data Analytics and Computational Intelligence (ICBDAC)*. 10.1109/ICBDACI.2017.8070809

Shrestha, A., & Mahmood, A. (2019). Review of Deep Learning Algorithms and Architectures. *IEEE Access: Practical Innovations, Open Solutions*, 7, 53040–53065. doi:10.1109/ACCESS.2019.2912200

Syed, J. N., Shree, K. S., Shurjeel, W., Mohammad, N., & Md. Asaduzz, A. (2019). Quantum Machine Learning for 6G Communication Networks: State-of-the-Art and Vision for the Future. *IEEE Access: Practical Innovations, Open Solutions*, 7, 46317–46350. doi:10.1109/ACCESS.2019.2909490

Vasileios, P.R., Sotirios, S., Panagiotis, S., Shaohua, W., George, K. K., & Sotirios K.G. (2021). Machine Learning in Beyond 5G/6G Networks—State-of-the-Art and Future Trends. *Electronics*, *10*, 2786. . doi:10.3390/electronics10222786

Yogesh, K., Komalpreet, K., & Gurpreet, S. (2020). Machine Learning Aspects and its Applications Towards Different Research Areas. *IEEE International Conference on Computation, Automation and Knowledge Management (ICCAKM)*. 10.1109/ICCAKM46823.2020.9051502

Zhao, Y., Zhai, W., Zhao, J., Zhang, T., Sun, S., Niyato, D., & Yan Lam, K. (2020). *A Survey of 6G Wireless Communications: Emerging Technologies*. arXiv:2004.08549v3.

Zhao, Y., Zhai, W., Zhao, J., Zhang, T., Sun, S., Niyato, D., & Yan Lam, K. (2021). *A Comprehensive Survey of 6G Wireless Communications*. arXiv:2101.03889v2.

List of Abbreviations

AMBC: Ambient Backscattering Communications
AI: Artificial Intelligence
APM: Amplitude Phase Modulation
CR: Cognitive Radio
DSA: Dynamic Spectrum Adaptation
DL: Deep Learning
DRL: Deep Reinforcement Learning
FL: Federated Learning
GAN: Generative Adversarial Networks
IM: Index Modulation
MDP: Markov Decision Process
ML: Machine Learning
NOMA: Non-Orthogonal Multiple Access
OFDMA: Orthogonal Frequency Division Multiple Access

QoS: Quality of Service
RKHS: Reproducing Kernel Hilbert Space
SM: Space Modulation
SSK: Space shift Keying
SWIPT: Simultaneous wireless information and Power Transfer
SR: Symbiotic Radio
SVM: Super Vector Machine
THz: Tera Hertz

Chapter 4
Security and Privacy Issues in the Internet of Things

Sridevi
Karnatak University, Dharwad, India

Apoorva Shripad Patil
Karnatak University, Dharwad, India

ABSTRACT

The internet of things describes the connection of distinctive embedded computing devices within the internet. It is the network of connection of physical things that has electronics that have been embedded within their architecture in order to sense and communicate to an external environment. IoT has turned up as a very powerful and promising technology, which brings up significant economic, social, and technical development. Meanwhile, it also brings up various security challenges. At present, nearly nine billion 'things' (physical objects) are connected to the internet. Security is the major concern nowadays as the risks have very high consequences. This chapter presents a detailed view on the internet of things and the advancements of various technologies like cloud, fog, edge computing, IoT architectures, along with various technologies used to prevent and resolve these security and privacy issues of IoT. Finally, future research opportunities and challenges are discussed.

INTRODUCTION

There have been a lot of technical breakthroughs which have made people to adapt to automated technology and make their living easier. Through the sensor network, IoTs collect a significant amount of data. By 2025, the IIoT is estimated

DOI: 10.4018/978-1-6684-3921-0.ch004

to produce "75.44 billion devices" and generate "79.4 zettabytes" data (Columbus, 2016). Because of additional sensors, higher computational power, and dependable communication, IoTs are rapidly increasing and will also increase in the upcoming years, which is shown in Figure 1. The increased use of IoTs has resulted in the unavoidable exchange of personal information, potentially leading to a data breach.

Figure 1. Increase in the number of IoT devices(Khanna & Kaur, 2019)

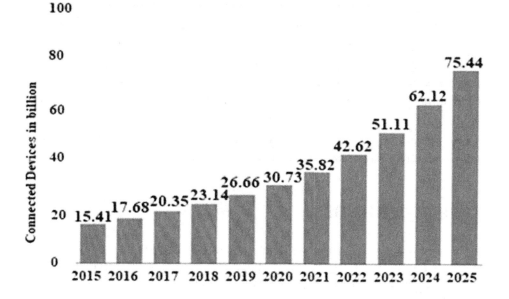

Data collection, Data transfer, and Data evaluation are the three important stages of the IoT system. The first stage i.e the data collection stage gathers data and helps in transmission, it includes sensor antennae and microcontrollers. The gateway and IoT hub or network, is utilized in the second stage where data transfer is done. The last stage has User interfaces and the back-end system, such as the cloud which is used for data analysis. Critical information from IoT devices will leak if there is a breach in these stages. Due to the lack of an effective encryption and authentication system, much vulnerability in IoT devices are recently emerged. Unfortunately, no complete security solutions to protect against security assaults are currently available in the market.

BACKGROUND

Evolution of IoT

A computer network, which was invented in the early 1960s, allowed two computers to communicate with each other. TCP/IP was introduced in the 1980s. In the 1980s, people began using the internet. Later, in 1991, the World Wide Web (WWW) became available, increasing the popularity of the internet. Business started happening online. The term "mobile internet" was introduced, where the mobile phones were connected to the internet. Through the internet the end users were connected via connecting devices as the result of their use of social networking. As indicated in Figure 2, the next stage in IoT is to connect all of the things in an environment and interact through the internet. The internet is connected to all smart objects in IoT. With the help of humans, they converse with each other. The IoT is quickly becoming the most popular and well-known IT concept. It draws users' attention by allowing them to connect from anyplace in the world and by giving each thing a unique identity (Aggarwal, 2012).

INTERNET OF THINGS

The technology is playing a very significant role in our lives today. Today, the Internet has widespread to everywhere. Internet has not only changed the lifestyle of the people but has also created a comfortable life. The way of our living has been greatly influenced by the introduction of new technologies. The discovery of internet has revolutionized the world. At present the whole world depends on the internet and computer to perform various operations.

The Internet of Things (IoT) is the interconnection of embedded physical things with the internet to collect and perform data exchange without any human intervention. Here the term "Things "means everything or anything in our day to day life that can be connected or accessed through the internet. IoT technologies bring up real time applications such as smart education, smart homes, smart transportation. IoT consists of many technologies such as Wi-Fi (Wireless Fidelity), Radio Frequency Identification Devices (RFID), GPUs (Graphics Processing Unit), Intelligent Sensors, Zigbee, etc. The IoT devices perform real time operations such as they collect the data, process it and also communicate with each other in real time.

This kind of communication which is real time is always subjected to connecting, monitoring and interacting with many other systems. These devices can store information such as light intensity, electricity consumption, geo location etc. IoT plays very important role in business by providing them real time insights, it shows

Figure 2. Evolution of Internet of Things (Gomathi et al., 2018)

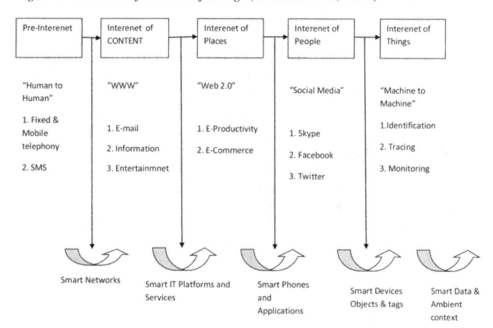

them how do their systems really look like, it provides insights into everything from machine's performance to supply chain, logistics etc. IoT helps in reducing the labour costs by automating the processes. It helps in providing transparency in customer transactions. By making manufacturing and delivery of goods less expensive, it improves service delivery. Generally, IoT is being largely used in the field of manufacturing, transportation, various utility organizations etc. It also has its use cases within organizations that are leading to digital transformation such as home automation industry, smart agriculture etc.

In an IoT environment all of the appliances and gadgets are connected to a network. IoT devices have embedded sensors, CPUs, actuators. The IoT is integrated with a set of technologies that work in unison. Actuators and Sensors enable users to interact with their surroundings. Sensors must be intelligently stored and analysed. Storage of data and its processing takes place at the remote server or on network's edge. If data preparation is done, it happens at the sensor or any of the devices nearby. After the data has been processed, it is then sent to a distant server. IoTs have limited storage and processing capabilities, due to size, energy, power, and computational capability constraints.

IoT systems are very complex in nature, hence providing security to IoT system which are vulnerable to attacks is very important. There are various security issues, some of them include:

- **Vulnerabilites:** These devices lack high computational capacity for providing built-in security and hence IoTs become vulnerable.
- **Malware**: It is a software designed to intentionally disrupt a computer server, client, network etc. The attacker gains unauthorized access to systems and leaks private information which hammers the user's security and privacy. Hence a strong security system is necessary.
- **Information theft and unknown theft**: These IoT devices are connected to the internet and therefore they have high chances of being exposed online which causes important technical and personal information leakage.
- **Device mismanagement and misconfiguration**: Poor password, security oversights and overall mismanagement can cause various security attacks. Lack of security knowledge among the users can also lead a way for their system being vulnerable to attacks. Service providers and manufactures must provide proper information about the system to users.

Some of the common practices to reduce risks and prevent threats are:

- Assign an administrator- Assigning an administrator can help in providing security to the IoT devices as be will be in charge of ensuring security of the device always and will also keep an eye on various operations which are being run on the system.
- Regularly check for patches and updates- Vulnerabilities can come from any IoT device layer. Some of the older vulnerability can also be used by the cybercriminals to infect the systems. Hence regular check for patches and required updates must be done.
- Use unique and very strong passwords – Strong passwords help to prevent cyberattacks as they cannot be guessed or revealed easily.

As IoTs have become a very important aspect in our life today a strong security system is necessary to prevent and protect these devices from the attacks.

Characteristics of IoT

- **Interconnectivity:** Almost all the devices can be associated with each other.
- **Heterogeneity:** IoT devices based on many different networks and hardware platforms, via different networks, IoTs communicate with other service or devices.
- **Dynamic changes**: IoTs states, like being connected or disconnected, also the speed and location, vary dynamically i.e. the number of devices is not fixed.

- **Massive scale**: As there are a lot of IoT devices the data generated is also more. It is very crucial to manage the data collected and to interpret for application purposes. Efficient data management and data semantics play a major role here.
- **Safety:** The safety of IoTs is an important issue for both recipients and makers of IoTs. The Security for the networks, the data and the end points is very important.
- **Connectivity:** It is through connectivity that the network compatibility and accessibility are possible. The ability to connect to a network is accessibility and the capacity to take in data and output data in a more common way is compatibility.

Applications of IoT

- **Smart Home:** They have become the revolutionary ladder to success, and it is expected that they will become as familiar as smartphones. These smart home products consume less energy and also save time. Some of the examples for the smart home firms are: Nest, Ecobee, Ring, August, etc.
- **Wearables:** These devices will have sensors and software within them to collect data from their users. Key user insights are derived by pre-processing the data obtained from the end user. These kind of gadgets are used in wide range of exercise, health, and entertainment needs. Some of the examples are Smart wrist watches, smart clothing, smart glasses etc.
- **Connected Cars:** It uses internet connectivity and onboard sensors to advance its maintenance, passenger satisfaction and its operations.
- **The Industrial Internet:** The Industrial Internet is nothing but the Industrial Internet of Things. To design smart machines software, big analytics, sensors make a major contribution. IIoT plays a very important role to provide sustainability for a long time and also provides quality control.
- **Smart Cities:** Smart city concept includes things such as providing a surveillance which is smart, transportation which is automated, efficient management of energy, water, advanced security etc.
- **IoT in agriculture:** Increasing the food production is very important and Government is providing various facilities to farmers by providing them very advanced scientific technologies. Ex: Some of the IoT applications in agriculture are soil moisture and nutrients sensing, efficient water management for plants and making decisions to select suitable fertliser.
- **Smart Retail:** IoT allows merchants to interact with their customers in order to improvise the in-store experience. Smartphones will allow shops to communicate with their customers at the times when they are not in the store.

Using Smartphones and Beacon technology to interact with customers can help retailers better serve their customers. They can also go after a customer's route through a store in order to improve the positioning of high-value items in high-traffic areas.

- **IoT in Healthcare:** Healthcare is still the Internet of Things' slumbering behemoth not only for businesses, but also for people's well-being, the idea of a connected healthcare system and medical equipments which are smart holds immense promise. The information gathered will aid in the personalization of a person's health and the development of tailored disease prevention programmes.

Advantages of IoT

- Users save a lot of time by automating processes.
- No matter how far the information is easily available and updated regularly in real time.
- It helps in safeguarding its users. If there are any dangers it can detect and alert the user. For example, GM OnStar – This gadget can detect a crash or a car accident on the road. It makes a call immediately if an accident or a crash takes.
- As IoT devices are connected, they interact with one another and conduct a lot of functions without human intervention thus reducing human labour. This can be a major benefit in manufacturing industry.
- In medical field IoTs play a major role. Patient's health can be monitored, helps in tracking wheelchairs, oxygen pumps etc.

Disadvantages of IoT

- As there is no completely secure and strong security mechanism for IoT, the systems become vulnerable to attacks and the sensitive data gets hacked.
- Without internet these devices cannot function.
- While completely being dependent on technology, users slowly will have no control over lives.
- As people depend on smart devices for everything they become lazy.
- Smart surveillance cameras, smart ironing systems and other technologies will take away jobs of innocent workers such as security guards, ironmen, and dry-cleaning services etc.
- It is extremely difficult to develop, construct, administer, and enable a large range of services.

IoT Architecture

There are different layers in the IoT architecture. Each layer's functioning is outlined below:

- **Smart device / Sensor layer:** Smart objects with sensors are there in the lowest layer. The sensors connects the digital and physical worlds, and collect and process real-time data. Sensors have memory, where in one can record a sum of measurements in some instances. It detects an attribute which is physical and convert it to instrument understandable signal. Sensors are being classified as per their purpose like body sensors, vehicle telematic sensors etc. Sensors in most of the cases have to be connected to the sensor gateways. This takes the shape of a Local Area Network (LAN) like Wi-Fi or Ethernet, or a Personal Area Network (PAN) like Bluetooth, ZigBee or Ultra Wideband (UWB). Wide Area Network (WAN) technologies like as GSM, GPRS, and LTE is used to connect sensors to backend servers/applications where in it is not necessary to connect to sensor aggregators. Wireless Sensor Networks are usually the sensors that connect using low-power and low-data-rate connections. WSNs support a lot more sensor nodes while still having great battery life and covering a lot of ground.
- **Network Layer-** It consists of networks and gateways. A very high sophisticated infrastructure is required as a medium for transmission. Current networks support Machine-to-machine (M2M) networks and their applications which have quite varied protocols. Different networks with various different technologies and protocols serve as a very important factor to work in heterogenous configuration to render support to large number of IoT applications and services. Latency, Security and bandwidth requirements are met with the help of networks which are public, private or hybrid.
- **Management Service Layer:** The IoT services are managed here. It also helps in securing analysis of information (Data Analytics), analysis of IoT devices and in device management.
- **Application Layer:** This layer provides a specific service by including all software which are required to provide the service. Here the data from the previous layer will be stored, aggregation will be done, the data will be processed. As a result, the data will be available to real time IoT devices (for ex smart watches).

CLOUD COMPUTING

The cloud computing technology is one of the sophisticated computing paradigm nowadays. According to (De Donno et al., 2019) Google and Amazon were the first to adopt the term "cloud computing" in 2006. Ref. (Elazhary, 2019) provided the most recent Cloud definition in 2019, as "Cloud is computer paradigm for providing anything as a service that is virtualized, pooled, shared, and can be produced and released quickly with low administration work".

Cloud includes three service models, they are Infrastructure as service (IAAS), platform as service (PAAS) and software as service (SAAS) (Hakak et al., 2013) and four different deployment models - they are "private cloud, public cloud, hybrid cloud and community cloud". Cost savings, scalability, efficiency are advantages of the cloud computing. There are also setbacks when dealing with large bulk of data like latency issues low latency will be there, security and privacy, high internet bandwidth, energy usage, analytics, load balancing, management of data. Furthermore, it is a centralized computing paradigm; the majority of computations take place directly in the Cloud.

FOG COMPUTING

Flavio Bonomi of CISCO was the first to mention and define the phrase "fog computing" in 2012. According to (Bonomi et al., 2014), fog is a virtual platform which provides various services such as storage, to compute and network services between the data centers and the end devices, as a result the end user's need is met locally rather than in the Cloud at a distance.

Furthermore, Fog Computing provides minimal latency and real-time processing. Fog layer is made up of the distributed fog nodes that are being installed on the the network's edge (Yousefpour et al., 2018). Each of the fog node has data storage, network communication facilities, onboard processing resources. By allowing the process of computation, networking,storage,management of data on network nodes close to IoTs, Fog Computing technology has recently been used to support a variety of applications, including smart grids, smart vehicles, smart homes (Yi et al., 2016), and smart agriculture (Hsu et al., 2020). For example, is an application where this Fog Computing technology provides high security, low cost, low latency and energy efficiency.

Some amount of data is being stored at the local data center in fog computing but this does not happen in cloud computing, this is one of the major differences between the fog computing technology and cloud computing technology. Along with this both of these technologies i.e the cloud computing and the fog computing consumes less energy and has lower operating expenses, real-time interaction, low latency, support for mobility, network bandwidth conservation, improved security, efficiency, and are also some of the distinctive features of Fog (Anawar et al., 2018); (Bouzarkouna et al., 2018). Farmers and agricultural stakeholders benefit from these distinct traits. The data which is being collected by all the devices, for example, the information which is sensitive that must be processed swiftly and locally, as a result, Fog Computing can give a benefit by allowing local processing and analysis to take place without having to send data back to the cloud as in cloud computing.

EDGE COMPUTING

Here the data is processed close to sensors or mobile devices. Cloud Computing technology confronts a some of the serious difficulties because of centralization. The edge computing technology has been designed to increase the cloud performance and the speed is also increased by allowing the processing of data and storage which is to be done locally at end devices. According to reference (Shi et al., 2016), this technology allows computation process to be done at network's edge. Heterogencity, context-awareness, low latency, location awareness, dense geographical distribution, closeness, mobility support separate Edge Computing from cloud (Khan et al., 2019).

In terms of virtualization service, high scalability, low latency, cheap bandwidth costs, mobility support, both edge computing and fog computing appear to be similar. Although edge computing has fewer resources, computing and storage capabilities, and is farther away from end devices than Fog it appears to be similar to the fog computing. (Hu.et.al 2019). Agricultural applications that benefit from edge computing include pest identification, agricultural product safety traceability, smart machinery used in agriculture, and intelligent management (Zhang et al., 2020).

ATTACKS ON IOT AND SYSTEM VULNERABILITIES

The Internet of Things encompasses a very huge number of devices and technology, from a very small and tiny embedded devices to massive servers. Hence there must be attention to security concerns at various IoT tiers. The perception layer, the network layer and the application layer are the layers that make up the most popular IoT layer architecture. Thera are various possible attacks that can happen at these layers, some of them are shown in Figure 3.

Table 1. Comparison of Cloud, Fog and Edge

Parameters	Cloud Computing	Fog Computing	Edge Computing
Scalability	Supported	Supported	Supported
Interoperability	Supported - Required capabilities for moving workloads between cloud providers and mitigating amongst providers when using services from various cloud providers	Supported-The components of the system can be distributed among multiple locations and they can be created by a variety of companies	Supported - As many suppliers and providers become more interested in IoT development, Hence strong interoperability is required.
Mobility	Not supported	Supported - In order to interact fog nodes and end devices need be able to move dynamically.	Highly Supported
Heterogeneity	Not Supported	Supported	Supported
Distribution across the globe	The cloud is, by definition, a distributed storage system, but it does not permit device deployment across geographical boundaries.	Decentralized and distributed deployment to suit fixed and mobile IoT/end-user devices is supported.	Supported
Location Awareness	Not supported	Supported	Supported
Performance	Congestion or server outages while processing can wreak havoc on cloud services, causing delays.	Fast response, low latency, and low bandwidth requirements are all supported.	Supports the fastest reaction time, most efficient processing, and the least amount of network demand, all of which improve performance.
Management of Quality of Service	Non-real-time processing is supported.	Real time Processing communication is supported.	Real-time support, with greater QoS (Quality of Service) and lower latency for end users.

Figure 3. IoT attacks

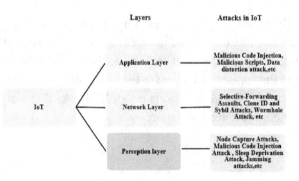

Perception layer security attacks

Objects with associated sensors, smart metres, robots, cameras, and other devices are included in the perception or device layer. This layer locates and collects target sensor data, such as movements, vibrations, atmospheric chemicals, heat, direction, humidity, etc. These data are delivered to the network layer, which subsequently sends them to a data processing system (Deogirikar, 2017). There are various types of attacks on perception layer:

- **Node Capture Attacks:** In this attack an attackers change the hardware, replace a node, capture a node. When the node is replaced then nodes show malicious behavior and the complete IoT network will not be secured(Sudeendra Kumar et al., 2018).
- **Malicious Code Injection Attack:** Once the malicious code is injected into the node's memory. These codes can be used to carry out nefarious actions and the entire network can be accessed by the attacker. Attacks are more common when software is updated over-the-air (OTA). An intruder may inject Trojans onto a device while it is in operation (for example, during a scheduled firmware update and requires device reboot). In this scenario, the security problem is categorized into two types: identification, correct authentication, of a network edge device, and ensuring that in the name of updates and upgrades no drivers or viruses are loaded on peripheral devices (Sudeendra Kumar et al., 2018).
- **Sleep Deprivation Attack:** These attacks are similar to denial-of-service assaults in that they completely drain the battery of the edge device, which is typically designed for low-power operation(Sudeendra Kumar et al., 2018).
- **Jamming attacks:** When this attack happens the communications are disrupted as the Signal - to-Inference-plus- Noise ratio (SINR) will be decreased by transmitting radio signals.

Network layer security attacks

This layer allows the processing and transmission of the data collected by the different devices and applications all the way through gateways or interfaces between diversified networks with the help various protocols. The various attacks on network layer are:

- **Selective-Forwarding Assaults:** These are distributed denial-of-service (DDoS) attacks. Malicious nodes forward just selected packets in order to interrupt the path's course, albeit any of the protocols would be targeted, as a result attacker may be able to forward all RPL (Routing Protocol for Low

Power and Lossy Networks) control messages while discarding other packets. Selective forwarding, when combined with a sinkhole or other assaults, can cause a great damage.

- **Clone ID and Sybil Attacks:** These attacks entail replicating the identity of a node (i.e. a valid node) to a second node for a lot of reasons, including gaining access to a larger area of a network. Sybil attacks make use of logical entities for a single physical node to govern a large network area without the usage of additional nodes.
- **Wormhole Attack:** Wormhole attacks are designed to attack "traffic flows" and "network typologies". These attacks are carried out by creating a tunnel that connects the two attackers and allows traffic to be selectively transmitted via this route. (Pongle & Chavan, 2015)
- **Denial of Service (DoS):** This assault disrupts a targeted network or a computation source, by lowering network capacity. DoS attacks against IoTs might take the form of Distributed Denial of Service (DDoS) or a basic DoS attack. The basic assault necessitates the use of a tool to send packets in order to crash a system, whereas DDoS can be carried out by a single attacker. They have the potential to disrupt and block network access. (Nawir et al., 2016)
- **Man in the Middle attacks:** These attacks use a variety of tactics to intercept and manipulate node-to-node communications. The attackers can watch the data once the node–node link is disrupted and real time manipulation of the data is done. (Nawir.et.al.2016)
- **Blackhole Attack:** Here, a malicious node is put within the network and discreetly drops all packets routed via it, passing nothing on. (Pongle & Chavan, 2015)

Application layer security attacks

In the architecture design the application layer is the last layer and is an interface between the IoT and the network with which the IoT communicates. This layer's security is a major difficulty, with a number of concerns that arise frequently, including the following.

- **Malicious Code Injection:** These assaults take use of coding within software to cause system harm or other undesirable effects. The anti-virus software will not be able to detect this malicious code. This malicious code can be activated on its own or it can be activated when certain action is performed by the user.(Abdul-Ghani et al., 2018).
- **Malicious Scripts:** these are the scripts that target networks or Internet-connected IoT devices. The attack is done by executing the codes that appear

to be normal codes but they actually are the malicious codes or x-scripts that steal data from the system and cause damage to the system (Gautam et al., 2019).

- **Data distortion attack:** this type of attack uses code embedded in software to harm systems or cause other undesirable effects. The anti-virus software will fail to detect these kind of attacks. The code gets activated by itself or when any action is being performed by the user.(Gautam.et.al.2019).

SECURITY IN IOT

IoT devices are now being used in a very large scale and these devices deal with very private and sensitive data. Because of this these devices are being attacked more often to access the sensitive data of the users. Hence to safe guard our network and the systems IoT intrusion detection systems must be created.

Intrusion Detection System (IDS)

An IDS detects suspicious behaviour and sends alerts when intruder found. IDS is a program that finds policy violations or malicious activities if any harmful activity or violation is found that will be reported to the administrator using Security Information and Event Management system (SIEM). Different sources a SIEM system collects many sources and also alerts filtering techniques to differentiate between false and malicious alarms. Intrusion detection systems are prone to false alarms. IDS products must be finely tuned when enterprises first implement them. There are five types of IDS.

- **Intrusion Detection System for Networks (NIDS):** In order to analyse the traffic which being set from all the connected devices, Network intrusion detection systems (NIDS) are being set up at a predetermined point in the network. It compares it to known threat database by monitoring all subnet traffic. The administrator will be issued an alarm whenever an assault has been found.
- **Intrusion Detection System for Hosts (HIDS):** These run on hosts or on separate devices. In this kind of IDS alerts are being sent to the administrator if any malicious activity is being found while monitoring the incoming and outgoing packets. Here the comparison is being done between the prior snapshot and the snapshot of existing system files. If the system files are being destroyed or modified the administrator will be given an alert if the analytical system files are destroyed or modified.

- **Intrusion Detection System based on Protocols (PIDS):** Here the system (or an agent) is always there at the server (at the front end of the server), which interprets and regulates the protocol between the server and the user. Web servers are being secured by these kind of IDS by regularly monitoring the HTTPS protocol stream and accepting the associated HTTP protocol. Until HTTPS reaches the presentation layer it is not encrypted, hence this system would have to stay in this interface in order to use HTTPS.
- **Intrusion Detection System Based on Application Protocols (APIDS):** It has the collection of servers. Intrusions are detected by monitoring and interpreting communication protocols that are application specific. For example, track the SQL protocol (There will be interaction between the middleware and the database present on the web server).
- **Intrusion Detection System (Hybrid):** Combination of two or more intrusion detection technologies forms the Hybrid IDS. Here the network information and system data is merged. This kind of IDS is more effective than any other IDS.

PRIVACY IN IOT

The current networks enable consumers to employ heterogeneous smart de vices to access various services like smart health services, automated vehicle services etc. The private data of the users is being utilized by various service providers when they deliver these services. Other stakeholders may be given access to the data and they use various advanced techniques to evaluate it and bring out new trends for their own product, which may be more fit for the market needs. In order to provide privacy for the user's data the service providers must provide proper information regarding how and foe what their data was used.

Most devices in the current network(5G) will depend on Location-Based Services (LBSs)(Liao et al., 2018). To provide services to users, an LBS leverages location data associated with the smartphone and/or mobile device. The promotion of LBS has recently increased significantly in a number of verticals, including healthcare, transportations, entertainment, government, delivery services and various others.

The LBSs make the life easier and more enjoyable, but they bring up certain privacy issue, such as being constantly tracked. Individuals will not have much information regarding the threats posed by these technologies. More crucially, digital media recently claimed that telecom firms are disclosing their users' actual location/position to a lot of stakeholders without their consent. As a result of this, LBSs may pose a risk to the privacy of its users.

- Identity Privacy- Identity privacy refers to the protection of a device's/ system's/identity-related user's information against active attacks. As more gadgets become connected to the Internet, the risk of identity theft rises alarmingly (Borgaonkar et al., 2015). For example, according to recent research, an active attacker can expose a subscriber's identity by collecting the International Mobile Subscriber Identity (IMSI) of the subscriber's UE. Furthermore, identity theft can reveal more information about a user. As a result, identity theft is one of the most significant concerns in the 5G and IoT industries. Designing an efficient and secure identity management techniques for identity privacy in 5G networks is critical.
- End-to-end data privacy: Current networks benefit a wide range of stakeholders, including service providers, operators, businesses, and new technologies in tandem with new business models.

To perform various operations such as storing, accessing and processing personal information from the customers the majority of these stakeholders rely on cloud computing. Consumers' personal data will be processed and shared by other parties for their own interests, posing a risk of privacy breaches. As a result, end-to-end data confidentiality must be considered to preserve consumers' privacy (Chen et al., 2016);(Li et al., 2018).

- Shared environment and loss of personal data ownership issues: The shared network infrastructure or virtual networks may be used in applications such as smart grids and heath care. Unauthorized data access and exchange may be a risk with such shared network infrastructures. As a result, strong and effective solutions that provide shared network infrastructure functionality without jeopardizing user's privacy are required.
- Different trust objectives issues: When the mobile operators and service providers(communication service providers) collaborate and they shift a section of their network to the cloud in a network, the stakeholders may have different trust objectives/priorities based on their own policies(Liyanage et al., 2018). They may not take into account all the aspects of data privacy for consumers.
- Issues in trans-border information flows: Personal data is very sensitive and important, but freely flows across borders due to global digitalization. When there is free flow of data, it is critical to have an individual or government authorization for data transfers, including how data is handled and it is being kept across borders.
- Issues with third parties: As application creators are frequently granted authorization to access the network, they may be able to expose or sell

personal information to third parties. For example, via mobile apps, "the health insurance portability and accountability act (HIPPA) authorises a share out of individual's health data." Furthermore, in a cloud network, the information sharing rule can cause substantial data-privacy difficulties. (Liyanage et al., 2018); (Aaronson, 2019).

Regulatory objectives to achieve privacy

- Global single market promotion and interest balance: "Global single market promotion" refers to all the applicable legislative practices or regulatory objectives that have to be encouraged to strengthen and enable global privacy policies free of regulatory barriers and internal boundaries.
- Encourage data portability: The data portability principle allows individuals and organizations to move their private information from one nation to another and one service provider to another, without having to adhere to legislated criteria (De Hert et al., 2018). As a result, in upcoming networks it is critical to promote data portability.
- Define global market privacy standards: To ensure interoperability and compatibility, data privacy regulations are most required in the context of a global market. Requirements for new privacy legislation must be set by different regulatory organizations around the world by interacting and cooperating with each other. For example, the EU-US Privacy Shield imposes obligations on US corporations to protect EU individuals' personal data.
- Encourage data accountability and stewardship: The current networks will involve multiple players, data accountability and responsibility will be essential. Responsibility must be taken up by various stakeholders to provide proper information regarding how and when they utilize the user's private data and should also take care of the rules that will be followed when the data is being accessed by other stake holders. As a result, all stakeholders must put in place meaningful and suitable safeguards to demonstrate accountability and responsibility for personal data.

RESEARCH CHALLENGES

IoT enabling technologies have been improved drastically but still there are issues that need to be addressed:

Privacy and Security IoT is the critical component of the internet with its rising applications, it demands the necessity to appropriately address trust functions and security. IoT is built on top of existing wireless sensor networks (WSN), it inherits the same privacy and security concerns that WSN does.

Various flaws and attacks in IoT systems have let us know the importance and the necessity for the highly comprehensive security designs that secure the systems completely. Many attacks take use of flaws in particular devices to obtain access to their systems and, as a result, make safe equipment vulnerable to attacks. This gap further stimulates complete security solutions, which include efficient research in applied cryptography for system and data security, non-cryptographic security techniques, and frameworks that assist developers in developing secure systems on heterogeneous devices.

More research on cryptographic security services that can run on resource restricted IoT devices is required. This allows a variety of experienced users to safely use and install IoT systems, despite the fact that practically all IoT devices have insufficient user interfaces. Additional areas such as communication secrecy, supplemental safety, message integrity, authenticy, trustworthiness, message integrity criteria should be included along with the protection and security components of the IoT. In commercial transactions, smart objects must be protected from aiding competitors' access to secret information stored in the devices and therefore exploiting their information.

In M2M (Machine to Machine) Communication most of the protocols are unable to provide adequate end-to-end reliability. The research can be done in framing protocols that ensure end-to-end reliability and also provides end-to-end congestion control.

CONCLUSION

The IoT is a technology that allows sensors and control systems to generate, exchange, and consume data without requiring human contact. IoT architecture, Cloud, Fog and Edge technologies, various security concerns at various IoT layers, Artificial Intelligence, Machine learning, Deep Learning, IDS, Security issues and privacy related information are presented in this chapter. It has also addressed layers of IoT: network layer, middleware layer, communication protocols, and application layer security concerns. It also gave a critical study of existing IoT solutions based on various security techniques, such as cryptography and intrusion detection systems (IDSs). The current state of IoT security and privacy as well as some future research directions to improve IoT security levels are described. It is designed to serve as a road map for improving security and privacy in IoT applications.

ACKNOWLEDGMENT

Ms. Apoorva Shripad Patil is thankful to the Karnataka Science and Technology Promotion Society (KSTePs) for supporting our work through the Ph.D. fellowship. [grant number-DST/KSTePS/Ph.D. Fellowship/PHY-04:2020-21/427.

REFERENCES

Aaronson, S. A. (2019). Data is different, and that's why the world needs a new approach to governing cross-border data flows. *Digital Policy, Regulation & Governance, 21*(5), 441–460. doi:10.1108/DPRG-03-2019-0021

Abdul-Ghani, H. A., Konstantas, D., & Mahyoub, M. (2018). A comprehensive IoT attacks survey based on a building-blocked reference model. *International Journal of Advanced Computer Science and Applications, 9*(3), 355–373. doi:10.14569/IJACSA.2018.090349

Aggarwal. (2012). *RFID Security in the Context of "Internet of Things."* First International Conference on Security of Internet of Things, Kerala, India.

Anawar, M. R., Wang, S., Azam Zia, M., Jadoon, A. K., Akram, U., & Raza, S. (2018). Fog Computing: An Overview of Big IoT Data Analytics. *Wireless Communications and Mobile Computing, 2018*, 1–22. Advance online publication. doi:10.1155/2018/7157192

Bonomi, Milito, P. N., & J. Z. (2014). Fog Computing: A Platform for Internet of Things and Analytics. *Studies in Computational Intelligence.*

Borgaonkar, R., Shaik, A., Asokan, N., Niemi, V. V., & Seifert, J.-P. (2015). *LTE and IMSI catcher myths.* Academic Press.

Bouzarkouna, I., Sahnoun, M., Sghaier, N., Baudry, D., & Gout, C. (2018). Challenges Facing the Industrial Implementation of Fog Computing. *Proceedings - 2018 IEEE 6th International Conference on Future Internet of Things and Cloud, FiCloud 2018*, 341–348. 10.1109/FiCloud.2018.00056

Chen, M., Qian, Y., Mao, S., Tang, W., & Yang, X. (2016). Software-Defined Mobile Networks Security. *Mobile Networks and Applications, 21*(5), 729–743. doi:10.100711036-015-0665-5

Columbus, L. (2016). *Roundup Of Internet Of Things Forecasts And Market Estimates.* Academic Press.

De Donno, M., Tange, K., & Dragoni, N. (2019). Foundations and Evolution of Modern Computing Paradigms: Cloud, IoT, Edge, and Fog. *IEEE Access: Practical Innovations, Open Solutions, 7,* 150936–150948. doi:10.1109/ACCESS.2019.2947652

De Hert, P., Papakonstantinou, V., Malgieri, G., Beslay, L., & Sanchez, I. (2018). The right to data portability in the GDPR: Towards user-centric interoperability of digital services. *Computer Law & Security Review, 34*(2), 193–203. doi:10.1016/j. clsr.2017.10.003

Deogirikar, J. (2017). Security Attacks inIoT. *Survey (London, England),* 32–37.

Elazhary, H. (2019). Internet of Things (IoT), mobile cloud, cloudlet, mobile IoT, IoT cloud, fog, mobile edge, and edge emerging computing paradigms: Disambiguation and research directions. *Journal of Network and Computer Applications, 128*(October), 105–140. doi:10.1016/j.jnca.2018.10.021

Gautam, S., Malik, A., Singh, N., & Kumar, S. (2019). Recent Advances and Countermeasures against Various Attacks in IoT Environment. *2nd International Conference on Signal Processing and Communication, ICSPC 2019 - Proceedings,* 315–319. 10.1109/ICSPC46172.2019.8976527

Gomathi, R. M., Krishna, G. H. S., Brumancia, E., & Dhas, Y. M. (2018). A Survey on IoT Technologies, Evolution and Architecture. *2nd International Conference on Computer, Communication, and Signal Processing: Special Focus on Technology and Innovation for Smart Environment, ICCCSP 2018,* 1–5. 10.1109/ ICCCSP.2018.8452820

Hakak, S. A., Latif, S., & Amin, G. (2013). A Review on Mobile Cloud Computing and Issues in it. *International Journal of Computers and Applications, 75*(11), 1–4. doi:10.5120/13152-0760

Hsu, T. C., Yang, H., Chung, Y. C., & Hsu, C. H. (2020). A Creative IoT agriculture platform for cloud fog computing. *Sustainable Computing: Informatics and Systems, 28,* 100285. doi:10.1016/j.suscom.2018.10.006

Khan, W. Z., Ahmed, E., Hakak, S., Yaqoob, I., & Ahmed, A. (2019). Edge computing: A survey. *Future Generation Computer Systems, 97,* 219–235. doi:10.1016/j. future.2019.02.050

Khanna, A., & Kaur, S. (2019). Evolution of Internet of Things (IoT) and its significant impact in the field of Precision Agriculture. *Computers and Electronics in Agriculture, 157*(January), 218–231. doi:10.1016/j.compag.2018.12.039

Li, S., Da Xu, L., & Zhao, S. (2018). 5G Internet of Things: A survey. *Journal of Industrial Information Integration, 10*, 1–9. doi:10.1016/j.jii.2018.01.005

Liao, D., Li, H., Sun, G., Zhang, M., & Chang, V. (2018). Location and trajectory privacy preservation in 5G-Enabled vehicle social network services. *Journal of Network and Computer Applications, 110*, 108–118. doi:10.1016/j.jnca.2018.02.002

Liyanage, M., Salo, J., Braeken, A., Kumar, T., Seneviratne, S., & Ylianttila, M. (2018). 5G Privacy: Scenarios and Solutions. *IEEE 5G World Forum, 5GWF 2018 - Conference Proceedings*, 197–203. 10.1109/5GWF.2018.8516981

Nawir, M., Amir, A., Yaakob, N., Lynn, O. B., & Engineering, C. (2011). 2014 2nd International Conference on Electronic Design, ICED 2014. *2014 2nd International Conference on Electronic Design, ICED 2014.*

Pongle, P., & Chavan, G. (2015). A survey: Attacks on RPL and 6LoWPAN in IoT. *2015 International Conference on Pervasive Computing: Advance Communication Technology and Application for Society, ICPC 2015.* 10.1109/PERVASIVE.2015.7087034

Shi, W., Cao, J., Zhang, Q., Li, Y., & Xu, L. (2016). Edge Computing: Vision and Challenges. *IEEE Internet of Things Journal, 3*(5), 637–646. doi:10.1109/JIOT.2016.2579198

Sudeendra Kumar, K., Sahoo, S., Mahapatra, A., Swain, A. K., & Mahapatra, K. K. (2018). Security enhancements to system on chip devices for IoT perception layer. *Proceedings - 2017 IEEE International Symposium on Nanoelectronic and Information Systems, INIS 2017,* 151–156. 10.1109/iNIS.2017.39

Yi, S., Hao, Z., Qin, Z., & Li, Q. (2016). Fog computing: Platform and applications. *Proceedings - 3rd Workshop on Hot Topics in Web Systems and Technologies, HotWeb 2015*, 73–78. 10.1109/HotWeb.2015.22

Yousefpour, A., Ishigaki, G., Gour, R., & Jue, J. P. (2018). On Reducing IoT Service Delay via Fog Offloading. *IEEE Internet of Things Journal, 5*(2), 998–1010. doi:10.1109/JIOT.2017.2788802

Zhang, X., Cao, Z., & Dong, W. (2020). Overview of Edge Computing in the Agricultural Internet of Things: Key Technologies, Applications, Challenges. *IEEE Access: Practical Innovations, Open Solutions, 8*, 141748–141761. doi:10.1109/ACCESS.2020.3013005

KEY TERMS AND DEFINITIONS

Artificial Intelligence: It is one of the major branches of computer science which is concerned on creating machines which are smart and intelligent. These machines perform various tasks that would require human intelligence.

Cloud Computing: Cloud computing is the process where various computing facilties such as servers, databases, analytics, networking, etc. are provided over the internet ("the cloud") to provide resources that are flexible, faster innovation etc.

Edge Computing: It is the distributed computer paradigm that brings up the data storage and the computation closer the place where the data is being generated.

Fog Computing: It is the decentralized infrastructure where the data, storage, compute are being located somewhere in between the cloud and the data source, but are not always, being placed near the network's edge.

Internet of Things: Through the internet these devices communicate with one another. The devices are nothing but the physical objects that have sensors, the processing power, software, and other technologies.

Intrusion Detection System (IDS): IDS generates alerts if there are any suspicious activity being carried out by continuously monitoring the system.

Privacy: It is the ability to control who can access the data on the system. Only the authorized users must be able to access the data.

Security: System must be secured by preventing it from being attacked by the unauthorized users hence ensuring the safety of the data in the system.

Chapter 5
Role of Artificial Intelligence in Cyber Security:
A Useful Overview

Vaishnavi Shukla
Vellore Institute of Technology, Chennai, India

Atharva Deshmukh
https://orcid.org/0000-0002-8039-3523
Terna Engineering College, Mumbai, India

Amit Kumar Tyagi
https://orcid.org/0000-0003-2657-8700
Vellore Institute of Technology, Chennai, India

ABSTRACT

Cyber security has been an emerging concern of individuals and organizations all over the globe. Although the increasing dependence of the world on the internet proves to be an advancement of technology, it also happens to be a threat to important and private information, monetary fraud leading to huge losses for organizations, and many other issues. In this day and age, cyber security has become a great necessity, and great efforts have been made to enhance it. Artificial intelligence has emerged to be of immense importance over the past decade and is expected to bear great fruit in the coming time. It is no surprise that organizations are depending on AI-driven technologies to protect their data. AI, along with concepts of machine learning and deep learning, is being used to develop new and improved means to help in the cyber security domain. The scope of this chapter covers a few artificial intelligence concepts which have been used in the past to ensure security, a few ideas which have been discussed for future implementation, threats of using AI, etc.

DOI: 10.4018/978-1-6684-3921-0.ch005

INTRODUCTION

History of Artificial Intelligence

We owe the seeding of the idea of Artificial Intelligence to Alan Turing who first gave a mathematical approach and idea in his paper Computing Machinery and Intelligence in 1950. However, computers in 1949 could not store commands but only execute, hence there was a need to fundamentally change computers. Also, computing was extremely expensive. These were the two major factors holding back Turing from following his pursuit.

It was then a program designed by Allen Newell, Cliff Shaw, and Herbert Simon called Logic Theorist which mimicked the problem-solving skills of a human and brought the idea of Artificial Intelligence closer to reality. This program is also considered to be one of the first programs of AI.

Even though with time the understanding of Artificial Intelligence grew the main issue was still not solved, the computers were not advanced enough to carry out intelligent programs. However, with time, the computational power of machines improved along with their storage capacity to a point where Moore's law had caught up. Moore's law states that the number of transistors on a microchip doubles every two years and the cost of a computer reduces by half. By the 2000s AI had achieved many of its goals and this was the starting of an era where Artificial Intelligence became a field everyone had their eyes on (Haenlein and Kaplan, 2019; Kaul, Enslin, and Gross, 2020).

History of Cyber Security

During a research project ARPANET, a program called Creeper was made which moved across ARPANET's network and left a message "I'm the creeper, catch me if you can" wherever it went. Ray Tomlinson-the inventor of email, made a program called Reaper which tracked and deleted Creeper. Reaper was the first instance of an Antivirus. By 1990s a lot of work on the antivirus part was being done and by 1992, the first antivirus program appeared. Soon as the world started coming online, virus attacks as well antivirus started getting more and more well known. Soon antivirus like McAfee came to the market, still being widely used. By the 2000s more antivirus came into the picture and now we can see an antivirus on almost every device (Warner, 2012).

CYBER SECURITY AND AI

Principles of Cyber Security

There are certain principles followed to ensure cyber security. They are as follows:

1. **Confidentiality:** The information should be secure and shared with authorized personnel only and no unauthorized parties should have access to the information
2. **Integrity:** Cybersecurity to ensure that the information should be accurate, consistent and free from any modification from unauthorized organization to maintain the integrity of the information
3. **Availability:** the cybersecurity efforts should not hinder the access of information by the authorized parties and also provide redundancy access in case of an outrage.

These are the basic principles of cybersecurity and we shall now look at the need of Artificial Intelligence in the field of cybersecurity (Štitilis, 2016).

AI in Cyber Space

An increase in the cybercrime activities initially posed a threat as there were not enough cybercrime professionals to handle the increasing threats. However soon after AI came to the rescue, making machines independent enough to handle these threats on their own.

Whenever a cyber activity takes place, AI plays its role in the following 3 stages-

Data is collected from the user-it is processed by the system being managed by the security vendor-the detection system flags the malicious activities and may or may not generate action in response. With newly emerging concepts like IoT (Internet of Things) it has come to the companies' notice that the cyber dependency of the users is going to serve as a platform for attackers to play their part.

There are two ways to reduce cyber-attacks in this case- slowing down the attackers and speeding up the defenders.

For speeding up defenders, AI has proven to be an excellent solution. Traditional security had to spend a lot of time manually processing the alerts and figuring out if they are harmful or not. With the growth in the AI sector, most of these tasks are being done by the machines independently and hence speeding up the defense.

In the past, the focus was first on categorizing the malware but nowadays companies are opting for models that don't look for individual pieces of malware; rather they look for the behaviour exhibited by the attackers. This behavior is then analyzed and the models are trained as per these instances. Hence it has become more common to opt for Machine Learning based threat detection systems. With time these models become more durable and eventually have the potential of detecting zero-day attacks (Rekha, 2020).

Increasing Cyber Threat

- Internet of Things is a fast-growing domain and we can see its ever-growing market of it. However, since it is dependent on the data of the user which is processed by the internet, any leak of this data can lead to significant risk. The main issue in this aspect is the lack of authentication and encryption of the data.
- Newly emerging payment systems like bitcoin and Ethereum have taken over the public eye. It is based on the concept of blockchain. Blockchain seems to have more applications in the coming years in the field of medical record management, decentralized access control and identity management (Deshmukh et al., 2022).
- Large connection of devices being infected by the malware is called Botnet. This is one of the most threatening attacks as the increasing use of IoT along with devices belonging to the same server, or connected to the same internet network is under threat. If the attackers get through and reach any one of these devices, it is not that great of a challenge to reach the other devices and if they happen to be lucky, with only one device's access they might be able to get a great deal of information and misuse it (Tyagi and Aghila, 2011).
- Studies have shown that android devices are one of the most targeted devices for malware attacks. Out of 14 applications made for malware detection, 8 were for android devices.

When an unauthorized party enters a network without the permission of the network owner, we refer to it as Network Intrusion.

Network Intrusion takes place in the following stages:

1. The information regarding the target is collected and analyzed for the weak and strong points. The information could be anything from email address to open-source details and specifically, any detail regarding the network is looked upon with great attention.
2. In the exploitation stage, the attackers patiently keep visiting the websites frequently visited by the victim. The attackers are usually slow and take their steps with utmost care to avoid getting caught.
3. The intruders then steadily increase their activities and start making their way into scripts and keys under the network.

4.   After this, there comes the main stage of installing the malware into the network starting with less harmful ones and making their way onto installing more powerful malware.

5.   Slowly the intruders gain complete control over the network and can now access any information they wish to. The intruders gather all the necessary information and then leave the network.

6.   Some of the intruders are also careful enough to erase their traces in order to prevent getting caught.

Different attacks of Network Intrusion include Trojans, Worms, Traffic Flooding etc.

A Network Intrusion Detection System analyzes incoming and outgoing packets to recognize patterns which may cause threat disguised as normal packets. For example, it can be installed on the subnet where firewalls are so that the NIDS can identify any attackers trying to break through the firewall. When a threat is discovered, the NIDS alerts the administrator which then overlooks the security of the network (Mishra and Tyagi, 2019).

MACHINE LEARNING APPROACH

Machine learning to a great extent is dependent on the data sets on which it is trained. The main task of NIDS boils down to classifying the packets as normal or abnormal. One of the most robust and functional methods which can be applied to perform this task is a supervised learning machine learning model. For a classification of normal or abnormal, or in another manner into 0(for normal) and 1(for abnormal or vice versa), we can opt for a binary classification model.

The model is first given a huge dataset having instances of real modern normal activities and attack behavior. This data is then used to perform feature selection, where only features correlated with label variable are selected. To perform this task, we can use any of the popular supervised learning algorithms. Here we have discussed three of them:

Logistic Regression

In this approach, the main prediction function comes out to be

y'=mx+c

Where,

y': predicted output

m: slope of the line

c: intercept

There are a number of methods and formulae which are used to calculate the value of m and c. These values are calculated after being trained by huge amounts of data. Once the best possible values of m and c are available, we then go ahead and use this function to test our data.

Support Vector Machine

The basic idea of a support vector machine is to draw a line or plane, or a hyperplane, which is farthest away from both datasets on a graph.

In the picture above we can see the green and blue data points, these belong to two different classes. A maximum margin hyperplane is a plane which is farthest away from the nearest data point of both classes.

We can further apply optimization techniques such as Minibatch gradient descent or stochastic gradient descent and get the best vector to classify our dataset.

When a new dataset is introduced or tested, depending on its location of it on the graph, it is classified into either of the two classes.

Decision Tree

A decision tree, as suggested uses a tree data structure and every internal node is a binary condition. It is drawn upside down with the root nodes at the top. It splits into two branches that meet another node; the nodes are yes-no branches. The nodes which do not split into branches are known as the decision nodes. These nodes decide which class the data belongs to.

All of these classification algorithms are used to classify our packets and for the detected abnormal packets, the system alerts the administrators who in turn check whether the network has been intruded or not and take the necessary actions.

Malware Detection

In the simplest terms, malware is a piece of code designed to harm computer networks and misuse the data in them. The trick however is that the malware only cause harm once they have been installed or implanted completely.

Malware detection is used to detect whether a certain program happens to be harmful or not. As discussed, since malware detection is also an act of classification, machine learning happens to be a great solution.

Figure 1. Support vector machine

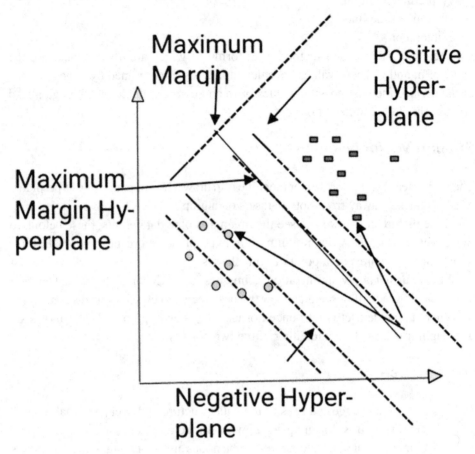

The growth of Neural Networks is a great step forward in the AI world. One of the biggest issues that AI has to overcome is false positives. False-positive refers to an error in a binary classification where an outcome wrongly indicates that an event has occurred when it actually has not. Neural networks improved the accuracy of models and also helped overcome false positives (Aslan and Samet, 2020).

An approach which is spoken of is Artificial Neural Network which is greatly used under the umbrella of deep learning. As discussed earlier, it is inspired by the neuron in the human brain and consists of multiple layers. The output of each layer is an input to the next layer. ANN is greatly used and is one of the best algorithms provided data is sufficiently provided. Although a bit more challenging to understand compared to the other algorithms, the output of ANNs is greatly appreciated in terms of the accuracy and robustness of the algorithm itself (Souri and Hosseini, 2018).

For malware detection, we again analyze the lines of code and normal codes are passed whereas abnormal codes are detected and analyzed further either by the AI itself or by cyber security personnel. On similar notes, AI has become one of the most important aspects of the cyber security domain in the past few years. Let us discuss a few benefits to understand what has caused this change.

BENEFITS AND SHORTCOMINGS

Benefits of Using AI and ML-Based Cybersecurity Models

- It improves over time. The Machine Learning and Deep Learning algorithms study and analyze the behaviours of business organizations and cluster them. It then detects any deviation in behaviour from the norm. With time, the data it analyzes makes the algorithms better and more sensitive.
- With time attackers have adapted to new technologies as well. The traditional methods may not be able to notice certain threats but AI has the potential to identify certain threats which may go unnoticed otherwise.
- On a daily basis, huge amounts of data are exchanged between networks and some of the malicious activities might go unnoticed by cybersecurity personnel. AI is able to process a huge amount of data and hence happens to be a better option.
- Analyzing and assessing the existing security measures through AI research can help in vulnerability management. AI helps you assess systems quicker than cybersecurity personnel, thereby increasing your problem-solving ability manifold. It identifies weak points in computer systems and business networks and helps businesses focus on important security tasks.
- An organization may have to deal with a phishing attack along with a denial-of-service attack or ransomware all at once. In these cases, prioritizing the attacks becomes essential as negligence at one spot can cost you a lot. AI helps in prioritizing these attacks and handling them using minimum time.
- Although avoiding major threats is a challenging task but regular basic checks must also always be made to avoid small attacks. These attacks might be neglected by cybersecurity personnel. AI on the other hand takes care of duplicative cybersecurity processes for basic security threats and prevents them on a regular basis.
- It accelerates the detection and response time. AI does regular system checks and keeps the detection up to date. Along with this its response time to an identified threat also keeps severe data leaks in check.

- AI ensures authentication every time a user logs into the website. These websites can contain information that is sensitive to the user and hence authentication becomes an essential step before accessing any website. AI uses the help of facial recognition, CAPTCHA, fingerprint scanner and other means to carry out regular authentications.

Is AI Always Good?

We have seen the advantage of AI in the current space but we cannot turn a blind eye to the increasing threats being caused by AI.

Newer attacks, different in nature from the once executed traditionally have been on the rise and these attacks are designed specifically to strike AI systems. AI researchers have agreed that AI has affected the cybersecurity landscape by expanding existing threats, introducing new threats, and altering the characteristics of cyber-attacks.

Existing Threats

In the current time, the concepts of AI and ML have become widely available to the general public and the availability of codes is also wide. This has gathered a large number of people understanding the working of malware.

The widening growth of cheaply available hardware and freely available software resources has resulted in an increase in the number of attackers.

Introduction to New Threats

The involvement of AI has not only accelerated the existing threats but also given birth to new methods of attacking.

One of the means has been termed "Deepfake". Using Deep Learning algorithms, machines have become capable of forging faces, voices, texts and many more such things. Certain algorithms are used where, on being fed thousands of images of two people, the system is able to break down their facial features on the basis of similarities and differences. This is used to beat facial recognition systems. On a similar note, voices too are forged. Some of these algorithms require less than four seconds of training audio to recreate human voices. One of the famous attacks on this track is when an AI based software imitated a chief executive's voice and requested a transfer of €220,000. The CEO thought that the chief executive was the one talking and transferred the amount from the German parent company to the Hungarian subsidiary (Tyagi and Sreenath, 2021).

Text manipulation is another dangerous threat. Although text manipulation is easy to beat but it can be used in various ways to waver the public opinion and can be used widely in political battles.

While internet and email scams have been around for decades, deepfakes have increased the tendency of these emails to appear to be real, trapping more and more victims.

Changes in Typical Characteristics

AI based cybersecurity is solely based on analyzing the lines of codes. The cyber attackers have thus opted for writing their codes in a way which exploits the AI vulnerabilities and makes the AI unable to function at its full potential.

Most of the cyberattacks are targeted at stealing information or disrupting the working of a system. Newer attacks tend to go forth with a long-term plan. They wait and study the working of the system and as time passes, begin to interfere with its working and steal the required information.

This can be done in many ways- giving flawed data to the models. The stained data set will interfere with the training of the models and then hamper the working.

Tampering the categorization models can be fatal to most ML models. Since the very basis of the ML models depends on the training data provided to it. If the algorithm itself is hampered, there is a chance that the categorization will prove to be ineffective and the cyber attackers will end up being successful in breaking through the AI system.

Ethical Challenges

AI systems are effective in self-healing and self-testing. However, without any human intervention, the machine can very well walk into a path of destruction. The independence of the AI system does reduce human efforts greatly but it also poses a threat to the ethical undertakings of a task which can be negatively impacted without any human supervision. The growing dependence of organizations on AI for cyber security could ultimately lead to the vast deskilling of experts and might eventually lead to a period when humans will be incapable of stopping the interaction between two AIs (Tyagi, 2016).

What the Future Holds

Back in the day, cybercriminals were people with good coding knowledge but that no longer is the case. Malware is sold around and bought by people making them capable of easily breaking into systems.

With automation also becoming one of the most popular fields of interest, malware have also started to become automated. In the future, malware will no longer be monitored by people. Malware and the security models will fight a battle of their own. Although this might sound like a mere fight between two programmed systems, the data at stake can cause huge losses to organizations and individuals.

As time goes by, cybercriminals have increased in number, especially considering the COVID phase where most of the industries shifted to an online mode of communication. But were the cybersecurity cells enough in number to counter it? The biggest issue in the coming future can be the shortage of enough skilled cybersecurity specialists which will help in maintaining the security of such huge amounts of data.

It is no surprise that most of the organizations are invested in stronger and better cybersecurity systems. The importance of these defense mechanisms is increasing day by day and will continue to grow in the coming future with Industry 4.0 spreading further and AI entering more and more industries.

While the threat of AI on jobs is a widely debated topic, cybersecurity employees have a bit of time before they start worrying about it. Although it will reduce the jobs in places where manual segregation of incoming data is required, it will create more jobs in areas where maintenance of the AI is required.

In fact, it has been found that the industry is lacking of expertise. This does create more job opportunities but it requires highly skilled individuals, hence there is a need to train more individuals in this field to fill the gap that has been created through the years.

CONCLUSION

AI has been in the market for a few years and its applications have been growing continuously. This has resulted in a huge change in industries. Every organization wishes to switch to an online mode of communication and have a more tech-friendly representation for themselves. While this approach does seem the most functional, cost-effective and attractive, it does provide a huge opportunity for the attackers to play their part.

The general public has also become more aware of the threat of cybersecurity and has become more cautious in terms of sharing their information online. While there have been efforts made by cybersecurity and Artificial Intelligence specialists to reduce the criminal activities going around, a huge difference will be made if the public itself is taught about the safe sharing of data and how to avoid websites and emails appearing to be malicious.

Most of people have antivirus and other programs installed into their systems to avoid getting into the hands of malicious attackers.

AI will improve with time and its use will grow further into more fields. As discussed earlier, although it will provide better security features, it will also provide a wider area of attack for cybercriminals.

REFERENCES

Aslan, Ö. A., & Samet, R. (2020). A comprehensive review on malware detection approaches. *IEEE Access: Practical Innovations, Open Solutions*, 8, 6249–6271. doi:10.1109/ACCESS.2019.2963724

Deshmukh, A., Sreenath, N., Tyagi, A. K., & Abhichandan, U. V. E. (2022, January). Blockchain Enabled Cyber Security: A Comprehensive Survey. In *2022 International Conference on Computer Communication and Informatics (ICCCI)* (pp. 1-6). IEEE. 10.1109/ICCCI54379.2022.9740843

Haenlein, M., & Kaplan, A. (2019). A brief history of artificial intelligence: On the past, present, and future of artificial intelligence. *California Management Review*, *61*(4), 5–14. doi:10.1177/0008125619864925

Kaul, V., Enslin, S., & Gross, S. A. (2020). History of artificial intelligence in medicine. *Gastrointestinal Endoscopy*, *92*(4), 807–812. doi:10.1016/j.gie.2020.06.040 PMID:32565184

Mishra, S., & Tyagi, A. K. (2019, December). Intrusion detection in Internet of Things (IoTs) based applications using blockchain technolgy. In *2019 Third International conference on I-SMAC (IoT in Social, Mobile, Analytics and Cloud)(I-SMAC)* (pp. 123-128). IEEE. 10.1109/I-SMAC47947.2019.9032557

Rekha, G., Malik, S., Tyagi, A. K., & Nair, M. M. (2020). Intrusion detection in cyber security: Role of machine learning and data mining in cyber security. *Advances in Science. Technology and Engineering Systems Journal*, *5*(3), 72–81. doi:10.25046/aj050310

Souri, A., & Hosseini, R. (2018). A state-of-the-art survey of malware detection approaches using data mining techniques. *Human-centric Computing and Information Sciences*, *8*(1), 1–22. doi:10.118613673-018-0125-x

Štitilis, D., Pakutinskas, P., Kinis, U., & Malinauskaitė, I. (2016). Concepts and principles of cyber security strategies. *Journal of Security & Sustainability Issues*, *6*(2), 197–210. doi:10.9770/jssi.2016.6.2(1)

Tyagi, A. K. (2016, March). Article: Cyber Physical Systems (CPSs) – Opportunities and Challenges for Improving Cyber Security. *International Journal of Computers and Applications, 137*(14), 19–27. doi:10.5120/ijca2016908877

Tyagi, A. K., & Aghila, G. (2011). A wide scale survey on botnet. *International Journal of Computers and Applications, 34*(9), 10–23.

Tyagi, A. K., & Sreenath, N. (2021). Cyber physical systems: Analyses, challenges and possible solutions. *Internet of Things and Cyber-Physical Systems*.

Warner, M. (2012). Cybersecurity: A pre-history. *Intelligence and National Security, 27*(5), 781–799. doi:10.1080/02684527.2012.708530

KEY TERMS AND DEFINITIONS

Cyber Security: Is the act of protecting computers, mobile devices, servers, networks, and data from malicious attacks.

Cyber Threat: Is on an increase especially threat to medical and financial data is severe and hence there is a need to tighten the security surrounding these fields.

Firewall: Is a network security device which is used to monitor the incoming and outgoing packets in a network. It follows a set of security rules, based on which it blocks the packets it finds suspicious.

Knowledge-Based Approach: Uses several IF-ELSE conditional statements.

Machine Learning: Is a branch of Artificial Intelligence which focus on using data and applying algorithms to make predictions and classify data. Machine Learning broadly has three types of classification algorithms, supervised learning, unsupervised learning and reinforcement learning.

Malware: Is a malicious software which is meant for damaging a user's device and is often installed into devices by unsolicited emails.

Neural Networks: It is inspired by the neural networks present in the humans and later led to Artificial Neural Networks (ANNs) which are used in Deep Learning. Neural Networks rely on data to learn and improve their accuracy but once the algorithm is finely tuned, it can be used to classify data at a very high speed.

Pattern-Based Approach: Data is fed into the algorithms and a pattern is noticed to carry out the algorithms.

Supervised Learning: Has data which already has a correct answer whereas, in unsupervised learning, the algorithms cluster the data without any prior knowledge. Reinforcement learning uses a penalty system where the algorithm rewards itself for correct classification and gives a penalty for incorrect one.

Chapter 6
Privacy and Security in Wireless Devices for the Internet of Things

Sridevi
Karnatak University, Dharwad, India

Manojkumar T. Kamble
Karnatak University, Dharwad, India

ABSTRACT

IoT devices are used to make human life easier and better by saving time and human energy. The IoT devices are controlled by an artificial intelligence system so that the devices can take the necessary decision and perform the work efficiently. The IoT devices are used in homes, offices, factories, schools, traffic signals, water supply management, power sector, security surveillance, hospital, vehicle monitoring, smart city, etc. IoT devices play lifesaving things in human life by continuously monitoring human health. In this chapter, the privacy and security-related IoT device issues were discussed with real-time attacks, and some counter-attacks have been explained. The flow of the chapter is organized as an introduction, architecture, functions, storage management, privacy, security, key elements of IoT, key technologies in IoT, and research opportunities in the IoT domain respectively.

INTRODUCTION

IoT is founded on the same principles as the human internet system. Human beings are connected by sharing things and information. In the same way, IoT is the collection

DOI: 10.4018/978-1-6684-3921-0.ch006

of devices that are interconnected through the internet. In earlier days, the IoT devices worked alone each device has particular instructions, and that IoT device perform that task. Because of this reason IoT devices use many external devices to be connected to a central server. So to reduce the number of external devices a new era has been started, in that all the IoT devices are connected remotely and share the information collected with the help of sensors. IoT devices are made up of software and hardware products with minimum size to place and use conveniently. IoT devices are mainly used for the collection and monitoring the environmental things. These devices can take command remotely from the user and also reflect on the command to perform the assigned task. This is possible because of the use of software and hardware things that are embedded in the IoT devices.

To maintain privacy and security, some software and hardware things are required in IoT wireless devices and those are given as follows.

Components of Wireless IoT Devices

1. **Actuators**: Actuators are system components that control the whole device and also it converts energy into motion.
2. **Sensors:** Sensors are software and hardware devices that are used in IoT devices to sense and collect environmental data. The sensors are the main components of any IoT device. There are many types of sensors are used in IoT devices like heat sensors, wet sensors, moisture sensors, touch sensors, motion sensors, etc. (Gupta, Khan and Sethi, 2019).
3. **Memory:** Memory is the foremost thing in any IoT device. The data which is collected by the sensors should be stored before sharing with any other device.
4. **RFID:** Radio frequency identification is used to uniquely identify the devices. It provides a unique id to each device in the network (Singh, Singh, and Saxena 2020).

Characteristics of Wireless IoT Devices

IoT devices are made with new technology so that they perform advanced things in less time and in an easy way so they appear unique. Some of the interesting characteristics of IoT devices are listed below

1. **Unique identity**: IoT devices use an identity management system to identify every device uniquely because a large number the IoT devices are produced and used in this world. While the IoT devices want to communicate with each other the device identity must be known previously. RFID i.e. radio frequency

identification is used for maintaining the uniqueness and providing a unique identity to IoT devices (Singh, Singh, and Saxena 2020).

2. **Dynamic nature:** The state of the IoT devices continuously changing with respect the environmental conditions. The devices are in the power-saving mode so because of this reason these devices get activated whenever necessary.

3. **Self-adapting:** The IoT device sensors keep changing their nature concerning environmental conditions. The devices have self-replication features so in all situation these IoT device provides the best information and results.

4. **Self-configuring:** The modern IoT devices are equipped with such a technology that they configure themselves like network configuration, the latest software updating, and all with minimum user intervention.

5. **Heterogeneity:** This is the highest level of updated thing in modern IoT devices. Different IoT devices with their different configuration, protocols, standards, platforms and different networks are embedded in modern IoT devices and still, provide the best and most reliable results. This is happening because of IoT's heterogeneity character.

Advantages of Wireless IoT Devices

1. **Efficient resource utilization:** The IoT devices utilize the resources efficiently which are connected to the IoT server. These devices keep inactive whenever they are needed and they suddenly get inactive when their usage has been completed.

2. **Time-saving:** The IoT devices save the required time by performing the work efficiently. These devices are equipped with a new technology system so that these devices can complete the work in time.

3. **Reduce human efforts and errors**: The devices are made to do the human work in an easy way like air quality, humidity, electricity, resource utilization, and management, and also reduces the human efforts because all these things are done only with these digital devices.

4. **Improved security:** The IoT devices are monitored and controlled remotely so inhuman absence also these devices keep working and providing the updates to their connected system.

5. **User friendly and easy to use:** The IoT devices are having a simplified user interface in such a way that any person can use them without any technical knowledge.

Disadvantages of Wireless IoT Devices

1. **Security:** Because IoT devices are wirelessly connected to a server database, any hacker or intruder can get into the network and disrupt the working environment.
2. **Privacy:** Privacy is the foremost thing in any domain and even in an IoT system. The privacy of the user can be stolen using a fake identity and using any other illegal access.
3. **Complexity:** The IoT devices are spread all over the world for human use and their network complexity, and maintenance also increase while providing private and secure communication models.
4. **Flexibility**: The devices are flexible with their nature and application so security should be provided all the way so that they couldn't get a compromise with privacy and security things.
5. **High cost and time delay**: The IoT devices are more costly as compared to any other electronic network connecting devices and also the devices take lots of time for communication and data collection (Alamer, 2021).

BACKGROUND

The things or devices in the IoT are connected in wireless technology that they communicate remotely and even share the data in the same way. Whenever a wireless term is considered privacy and security get important automatically. Privacy is to maintain personal things within a shared group and security is to maintain the privacy of a particular group. To secure the IoT ecosystem (Gupta, Khan and Sethi, 2019), several security techniques have been proposed. Reducing delay and boosting the computation speed of security mechanisms in IoT systems is a challenging job in the real world. The bulk of Internet of things applications are managed by smartphone applications, which makes security a big concern. To safeguard data from attackers, several lightweight cryptographic algorithms have been devised. By simplifying key generation procedures, hybrid encryption algorithms provide strong security while simultaneously boosting computing performance (Swamy and Kota, 2020). In IoT networks, to decrease data analysis and transmission delay, many middleware technologies have been created. Permission, confidentiality, cryptography, secured network interfaces, and adequate program security are the main security issues in IoT middleware.

The Internet of Things (IoT) is organizing its network itself so that it joins any node at any time. To improve the quality of services to genuine devices, unique privacy-preserving lightweight device authentication systems must be designed and developed. The use of biological traits to authenticate individuals is becoming more popular, which creates new issues. In the field of security, the use of machine learning algorithms to forecast DoS and DDoS attacks is becoming more common (Swamy and Kota, 2020). To construct efficient security solutions, these algorithms require legitimate IoT traffic information. Since the majority of communication is dependent on wireless standards, radio jamming is one of the key difficulties with IoT implementation. By transmitting a strong radio close to an IoT device, an attacker can impede data transmission. Because of the usage of predictable smaller secret keys, Bluetooth low-energy devices in IoT systems are vulnerable to numerous assaults. Blockchain technology has an impact on supply chains, Industry 4.0, and power generation (Swamy and Kota, 2020). The blockchain's decentralized architecture, on the other hand, leads to increased energy consumption, storage, and computing overhead. The resource-constrained in IoT devices and energy-efficient blockchain solutions must be created Mohanta et al., 2020), because of its network design and varied access, 5G creates several security concerns for IoT communication in the future. Algorithms for verification, authentication and authorization, trust frameworks, and data privacy protection are being designed and developed to improve security in 5G IoT networks (Swamy and Kota, 2020). The wireless devices use cloud storage using blockchain technology for data storage and in (Goyat et al., 2020) terms of data communication the detection accuracy, certification latency, and computing overheads, the author achieved better results. The algorithm delivers 19.33 percent better outcomes in terms of average detection accuracy, according to the simulated results and comparison analysis. The paper's trustworthiness and efficacy were assured by storing a big amount of data on the cloud.

The eves dropping plays a very important role in wireless IoT devices while transmitting the data from one device to another. To make a secure connection in multiple antennas (Anajemba et al., 2022), the author used optimal jamming parameters for secure data communication in full-duplex communication and a stochastic optimization algorithm for improving privacy capacity in IoT multiuser scheme. In 2022 (Vakhter et al., 2022), the author discovered knowledge on threat modeling applicable to tiny wireless biomedical devices is reviewed and summarised in this study (MWBDs). The author then employed a domain-specific qualitative-quantitative threat model to assist MWBD designers and manufacturers in identifying risks and incorporating security into their designs during the pre-market phase of an MWBD's lifespan. In this chapter, It also emphasizes non-invasive direct assaults on telemetry interfaces in this study article. The results of the risk analysis conducted in this study show that the model is simple to use and sufficient in revealing dangers.

The Architecture of Wireless Devices

Wireless devices are electronic devices which are connected to other devices without physical channels and controlled with the help of the internet are called wireless devices. The wireless devices are meant for data collection and data sharing using internet technology. Figure 1. Shows the basic structure of how wireless devices are connected.

- **BSS**: The Basic Service Set (BSS) is a collection of all PHY layer stations that may communicate with one another. The BSSID, which is the MAC address of the access point that services the BSS, is assigned to each BSS. The basic service set is the fundamental component of an 802.11 wireless LAN. A BSS is a single access point (AP) that includes all linked stations in infrastructure mode. Not to be confused with the basic service area, which is the area covered by an access point (BSA). The access point controls the stations inside the BSS like a master. One access point and one station make up the most basic BSS (Github, 2022).
- **ESS**: An Extended Service Set (ESS) is a collection of two or more interconnected wireless BSSs that share the same SSID (network name), security credentials, and integrated wired local area networks that appear as a single BSS to the logical link control layer at any station associated with one of those BSSs, allowing for mobile IP and fast secure roaming applications. The BSSs may operate on the same or separate channels. An ESS's access points are connected by a distribution system. Each ESS is issued a 32-byte character string called an SSID (Github, 2022).
- **AP**: In a wireless network, an access point (AP) is a stationary device that operates as a base station. It facilitates station communication and access. If necessary, it also serves as a bridge between the WLAN and the wired network. APs are often deployed as standalone devices (Github, 2022).

The Layers in the IoT System

The IoT devices are organized into three layers. The data which has been collected by IoT sensors has to pass through three layers to get reach other devices or central databases. The three-layer architecture is shown below in figure 2.

1. **Physical layer:** The physical layer is the lowest in the architecture of the IoT system. The physical layer or perception layer is responsible for collecting the data from the environment using sensors. This layer consists of hardware and

Figure 1. Architecture of wireless devices (Github, 2022).

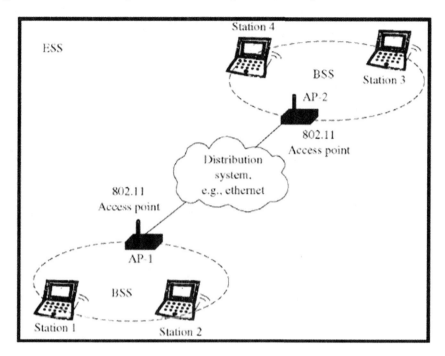

software devices that are used for sensing and collecting information or data. The following are the IoT physical layer attack

2. **Network layer:** This layer is responsible for transmitting the collected data from the physical layer to the memory and application layer. In this layer, the data is flowing from one sensor to another and the privacy of the data may get compromised by many attacks and vulnerable issues.

3. **Application layer:** The application layer is the end-user layer or device layer. In this, the device-to-device communication is carried out to exchange the data or information collected by the IoT sensors. There are many privacy and security issues are there in the application layer.

Figure 2. Layers in IoT system (Hameed and Alomary 2019).

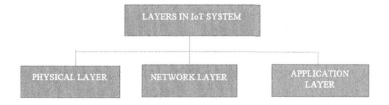

Figure 3. Attacks on IoT devices (Hameed and Alomary 2019)

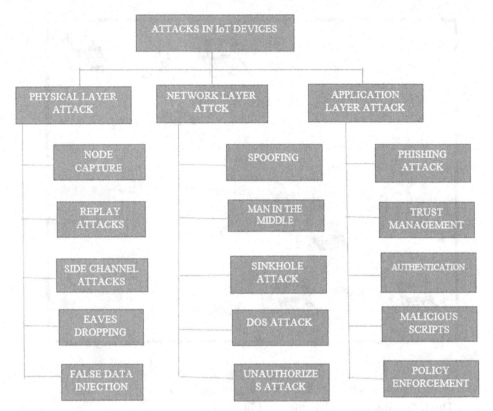

Attacks on Wireless IoT Devices

The IoT devices are facing many attacks in their environment. Some of the attacks on IoT devices are shown in figure 3.

Physical Layer Attacks

1. **Node capture attack:** As a result of the IoT applications, intelligent systems are placed in various places. To get network access, the attacker can either acquire the devices or swap them with the incorrect device. It's difficult to tell the difference between a smart device and a fake device in a physical attack. The intruder may obtain critical relevant details using this sort of attack. This sort of assault must be addressed to make the network more secure (Hameed and Alomary, 2019).

2. **Replay attack:** In a replay assault, the intruders steal data from the transmission device and broadcast the same information to the channel. An attacker in an

IoT ecosystem can hijack smart devices and transfer data as if they were an allowed device in the connecting group (Hameed and Alomary, 2019).

3. **Side-channel attack:** In a side-channel attack, the hackers will try to extract plain text from the encrypted message. Because the majority of encryption systems need key swaps for both encryption and decryption. Acquiring the key via a time constant is critical in this approach (Hameed and Alomary, 2019).

4. **Eavesdropping:** Through data transmission between two or more nodes in a network, an unauthorized user can get access to information. To access the data between the sender and the receiver, the attackers take advantage of insecure network connectivity. The data is not altered in this type of attack, but privacy is jeopardized (Singh, Singh, and Saxena 2020.; Hameed and Alomary, 2019).

5. **False data injection:** Devices or smart gadgets are placed at various places to gather data from the environment. Sensors and intelligent gadgets are capable of collecting data and passing it on to another layer. Smart gadgets are vulnerable to attackers due to their constrained resources. The attacker attempts to seize the device or extract the value from an insecure transmission medium to insert fake packets into the system (Hameed and Alomary, 2019).

Network Layer Attack

1. **Spoofing:** The attacker tries to get access to digital devices at the network level. After obtaining control of the nodes, the intruder acts as if it were a real network node. The network is inundated with fraudulent communications (Hameed and Alomary, 2019).

2. **Man in the middle attack:** This attack is very dangerous in wireless networks. In this, a hacker hacks the network and keeps an eye on the movement like data sharing between the devices. In this way, the person or hacker can get all the information flowing between devices and that person can manipulate that data too for his benefit.

3. **Sinkhole attack:** The Sinkhole attack has the potential to disrupt the internet of things' whole communication system. It diverts traffic via bogus routing, interrupting the routing flow. This might cause the network's whole traffic to be disrupted. One of the sinkhole attack approaches is to target the regular node first, then imitate the shortest way. The regular nodes then use this path to route their packets. As a result, packets are only partially or completely received at the needed destination node. The performance, efficiency, and dependability of network communication are all harmed by sinkhole attacks (Singh, Singh, and Saxena 2020).

4. **Denial of service (DOS):** In this method, the attacker sends unnecessary garbage requests to the server and the server gets down because of garbage

requests. When an authorized user wants to access and use the server using the internet of things then the server denies the user's request. This process is called denial of service.

5.   **Unauthorized access:** The attackers are after devices with limited resources that are connected to Enabling technologies. Because the majority of the devices were linked through a separate gateway. Various techniques are used by the attacker to acquire authorization information. After obtaining the private identity, the attacker gains access to network data (Hameed and Alomary, 2019).

Application Layer Attacks

1.   **Phishing attack:** A phishing attack occurs when someone tries to trick you into supplying confidential information over the internet. Phishing is frequently brought out using email, adverts, or internet sites that look to be identical to the sites you already use (Swamy and Kota, 2020).

2.   **Trust management:** The challenge of dealing with a trust-based issue on the application layer is difficult. As an IoT-related application for real-time monitoring and management of the environment, users can exchange private data across the network. In a decentralized setting, data is communicated and published to the network during computation. As a result, trust governance concerns will occur among the network's nodes. If a node in the network reacts unlawfully, the system should detect it. As a result, in the IoT system, adequate trust management is critical (Hameed and Alomary, 2019).

3.   **Authentication:** Smart gadgets, sensors, controllers, and certain electronic objects make up an IoT application, which monitors and computes. Smart gadgets take data or information and pass it over to the next level for analysis and processing. The network node sends the related event after the calculation is completed. Actual data from the sensors is essential for secure and efficient computing. The system becomes corrupt if the attacker acquires the sensors or smart objects. Every device on the network must be registered or authenticated to do this. One of the most crucial aspects of any IoT platform is authentication (Hameed and Alomary, 2019).

4.   **Malicious attack:** Because of unsecured communication routes and wireless connections, smart gadgets in IoT applications are exposed to the outside world. An attacker can use an application to inject malicious code into a device, putting it at risk of being hacked (Hameed and Alomary, 2019).

5.   **Policy enforcement:** Policy enforcement refers to applying some rules and regulations to access IoT devices. The enforcement is to maintain the privacy

and security concerns to the IoT devices and to the user who wants to use the system (Hameed and Alomary, 2019).

INFORMATION STORAGE SYSTEM IN WIRELESS DEVICES: SECURITY ISSUES

Centralized Storage System

All IoT devices are connected to a central server in this system, which collects and stores the acquired data. The centralized storage system has created a single network and it binds all the devices for communication and data interchange. The storage system has an advantage over the less use of devices that are used in connecting IoT devices in the network. This system suffers from many disadvantages like security issues:

- IF the hacker or intruder hacks any device which is connected to the central server then all IoT devices and even the central server get compromised.
- When the centralized server gets down then all the devices can't share their data with the server and other connected devices.
- When the centralized server gets too many requests and responses then other devices should wait until it gets free for data sharing.
- The host who maintains the central storage system can monitor or even disclose the private data to others. This is one of the main privacy issues in the centralized storage system.

Decentralized Storage System

In this system, the IoT devices are clustered and for each cluster, a storage server has been created so that all IoT devices are not dependent on the single storage system. The Decentralized storage system provides multiple storage systems for multiple devices so no other device waits for exchanging data between node and server. This system uses the block-chain concept that the data are distributed in small chunks and distributed over multiple computers. In this system, the network administrator has to maintain and secure all the distributed storage systems at the same time. Like this, the data is divided and stored in multiple devices except in a single central storage.

COMMUNICATION STANDARDS IN WIRELESS IOT DEVICES

IoT devices can use wired or wireless methods to communicate. Wireless norms of communication are becoming more widely used in society. As a result, this study describes wireless communication techniques over short and long distances. that are utilized in the Internet of Things. At the moment, major organizations such as IEEE, Internet Engineering Task Force, and Internet Engineering (Swamy and Kota, 2020). In general, the perception layer is a collection of sensors that collect and transmit physiological signals to the gateway using short-range communication protocols. The perception layer's sensors are mostly battery-powered and send data using short-range, low-power wireless communication technologies such as RFID, BLE, WiFi, ZigBee, Zwave, Thread, and Light-Fidelity (LiFi), and Wireless HART (Swamy and Kota, 2020). The usage of these communication protocols will extend the sensor's life (Swamy and Kota, 2020). Gateways frequently employ long-range communication technologies to exchange data between faraway places. For long-distance communication, LoRaWAN, NarrowBand-IoT, and Sigfox are often utilized, while 5G technology is slowly gaining traction. These standards offer benefits such as higher data rates, lower power consumption, and a larger service (Swamy and Kota, 2020).

RFID

Radio frequency identification is a technology for wireless communication that allows things to be automatically identified, tracked, and data collected Narasimha Swamy & S.R.Kota.2020). RFID is made up of RFID tags and RFID readers. RFID-tag is a microchip that stores identifying information and then uses that information to monitor items afterward. When an RFID tag comes into touch with an RFID reader, the data is retrieved. RFID readers can be stationary or portable. RFID tags have a communication range of 3 to 90 meters. Production monitoring and control, supply chain management, and other applications benefit from this form of communication (Swamy and Kota, 2020).

WIFI

802.11 radio technology is used in WiFi to transmit and receive data over short distances. There are several types of WiFi on the market now, including 802.11n, 802.11g, 802.11a, and 802.11b. The wireless fidelity frequency is around fifty meters, and the data throughput is up to 11 Mbps. WiFi may be found in smart homes, campuses, and other places (Swamy and Kota, 2020).

Bluetooth Low Energy

It's also called the Bluetooth Smart, and it's designed for short-range communication. Bluetooth low energy uses less power and enables higher transmission rates for audio and video applications than traditional Bluetooth technology (Swamy and Kota, 2020). It operates on the 2.4 GHz ISM band and has a range of up to 100 meters. BLE has a data throughput of up to 1Mbps. Bluetooth low energy is useful for applications such as smart homes, smart workplaces, smart retail, and so on (Swamy and Kota, 2020).

ZigBee

The ZigBee Alliance established a communication standard for carrying tiny amounts of data. ZigBee is a low-power, low-cost wireless technology that uses less than 1 milliwatt most of the time. The ZigBee's operational range is up to 150 meters. The channel's bandwidth is 1 MHz, and the data transfer rate is around 250 kbps. ZigBee is particularly beneficial in smart home and industrial 4.0 applications (Swamy and Kota, 2020).

Z-Wave

This Zensys-developed communication protocol establishes communication over a small range. To make it possible for node-to-node communication, Z-Wave creates mesh networks. When delivering smaller packets at a rate of 100kbps, Z-Wave minimizes delay. This technique has a transmission distance of around 30 meters (Swamy and Kota, 2020).

Light-Fidelity

LiFi is a wireless connectivity protocol that sends data via the visible light spectrum. This standard is sometimes referred to as Visible Light Communication. LiFi enables more mobility and access for many users. It has a data rate that is a hundred times quicker than WiFi. The range of communication, on the other hand, is quite limited. This standard benefits smart homes and Industry 4.0, as well as virtual and augmented reality applications (Swamy and Kota, 2020)

Wireless HART

Wireless HART is a more enhanced form of the wired HART protocol that is widely used in automated processes. In industrial networks, this standard allows for data exchange, self-organization, and self-repairing network architecture. The 2.4 GHz ISM frequency is used by Wireless HART. The protocol's data rate is 250 Kbps, and it spans a radius of 230 meters. Factory automation 4.0 applications are best served by this protocol(Narasimha Swamy. et al.2020).

PRIVACY AND SECURITY OF WIRELESS DEVICES IN THE INTERNET OF THINGS

The IoT's dynamic nature in terms of technology and functionalities, as well as the development of better techniques for communicating with the IoT, has resulted in distinct privacy threats and challenges. Privacy and security are very important concepts in IoT. The privacy is to access the IoT device by an authorized user only. The security is to maintain privacy for each IoT device because these devices have low energy, processing capability, and compute, secure communication among the devices participating in IoT is necessary (Tank et al., 2016). Many IoT devices' privacy gets compromised because of many reasons like weak authentication, communication without secure encryption, shared Wi-Fi network, network vulnerability, less featured cheap devices, etc. Small IoT devices are connected to a large network and small devices are made up of fewer security features so because of this reason a large group of secured devices also get compromised. The European general data protection system (GDPR) tells that personally identifiable information should be protected on the internet (Tank et al., 2016). Wireless sensor networks compromise more easily than a wired network.

Several privacy measures were offered in the field of privacy protection. The purpose of any confidentiality measure is to determine how much an intruder potentially violates a user's privacy. There are two levels of privacy metrics that have been identified: Information privacy metrics and geospatial confidentiality metrics. The most widely used privacy metric is location entropy. It can be used to determine the confidentiality of user geolocation by evaluating the variety of alternatives accessible, as well as the complexity of recognizing a participant's specific preferences, and specifications.

Initially, this metric was inspired by Shannon's entropy in information theory. It's used to figure out how susceptible region information is in Geospatial Based Server queries by estimating how much information an attacker can get by one of the region updates. Several academics proposed new privacy metrics based on the concept of location entropy. The authors defined location entropy as the availability of a geographic area. The attractiveness of a public area is determined by the entropy of visitors' footprints within it. Researchers are using location entropy to generate new privacy measures once again. It shows a stronger opponent who may utilize recorded data to DE anonymize fake identities more accurately, the entropy prediction depicts a more doubtful image in which the client has less privacy. K-anonymity is a further well-known statistic for location query privacy protection. The concept of k-anonymity was first introduced in the database research sector, and it quickly became a widely used privacy metric in the location privacy protection community (Abuladel et al., 2020). In this case, the k-anonymity measure means the condition in which the location information provided in an LBS inquiry pertains to a region where the question guarantee is undefined from at least (k–1) different clients also present nearby. Some academics used the idea of k-anonymity to produce better privacy indicators based on the creation of an improved anonymity set. The central idea makes use of the number of customers in a given location to convey the k-anonymity level's quality. When users are present throughout a region, ubiquity is established, extending the anonymity of the total location (Abuladel et al., 2020).

The protection of personal information matrix is the quantity of information an enemy can gain is measured by Information Gain or Loss, with the assumption that the better the privacy, the little information an opponent can get. Several data gain assessments, like uncertainty metrics, are based on data hypotheses. Information gain measures, on the other hand, explicitly consider the amount of preceding data. The qualities of conspicuous or disseminated information are measured using data similarity measures. They are usually independent of the opponent and assess the privacy degree solely based on the characteristics of released data. In practice, these metrics begin in the database domain, where they are most commonly used for information disinfection and distribution. Adversary's Success Probability When measuring privacy, the adversary's success probability shows how likely the adversary is to succeed in a single attempt, or how often they would succeed in a series of tries. Low achievement probability is associated with a high level of privacy. While this presumption remains true for a client population with a median value, an individual client may lose protection in any circumstance where the adversary's chances of victory are minimal. Error-based metrics, for example, use the difference between the true result and the gauge to determine how accurate the adversary's estimate is. Low privacy is associated with high correctness and few faults(Amani Abuladel. et al.2020).

The wireless sensors have created a network or group of sensor devices for exchanging data between the devices and servers. Some of the wireless sensor network attacks are discussed below

Distributed Denial of Service (DDOS)

DDOS is the extension of DOS. In DDOS the attacker sends the unnecessary garbage request to the server with multiple IoT devices so that the server gets down faster than the DOS technique.

Unauthorized Access

The Unauthorized access can also call it as an invalid user. The attacker gets the credentials of a valid user by hacking his system and logging into the IoT server with the help of an authorized user's credentials.

Node Replication Attack

A node replication attack is an attempt to introduce one or more nodes to a network with the same ID as another node. As a result, the new node can cause duplicate entries, erroneous data packet routing, inaccurate sensor readings, or network disconnection, compromising the functionality of a sensor network(A. Hameed.2019).

Selective Forwarding Attack

As the name implies, packets are forwarded selectively rather than as intended in selective forwarding. A malicious node is mostly responsible for this selection. This network node drops packets rather than forwards them. As a result, data transmission is corrupted. It is quite difficult to locate a rogue node in such circumstances. These nodes not only drop packets but also send them to the wrong recipient, causing the information or data to be corrupted(A. Hameed.2019).

Sybil Attack

These exploits first surfaced in the setting of a peer-to-peer network. It poses a significant danger to WSNs. The Sybil attack generates warped identities for valid nodes. Identity theft can affect anywhere from a few to a large number of nodes, interrupting a large percentage of the network. As a result, network security features like shared storage, inequality, and multipath become less effective(A. Hameed.2019).

Information Manipulation

This security issue will arise when an unauthorized person accesses the server and start manipulating or changing the information which is stored in the server database.

This can be done by a man-in-the-middle attack that accesses the data which flows between the wireless devices. Information manipulation can be also possible by an identity theft attack in that the hacker gets the authorized user data by hacking the user system and logging into the database server as an authorized user and starting to manipulate the data from the database.

Information Disclosure

Information stored in a database server may be disclosed whenever an invalid user succeeds in using another server by hacking the user credentials.

Localization and Monitoring

Localization and tracking is the threat of determining and documenting an individual's location through reality. To relate persistent constraints to one person, tracking necessitates differentiating proof or something to that effect. Tracking is now possible using a variety of ways, including the Global Positioning System, online traffic, and cell phone location. Many tangible privacy violations have been identified as a result of this threat(Amani Abuladel. et al.2020), such as GPS stalking, the exposure of private information such as an illness, or the general unease of being watched. Individual monitoring and location, on the other hand, is a significant feature in many IoT systems(Amani Abuladel. et al.2020). Clients regard it as an infringement when they don't have control over their location data, are unaware of its disclosure, or when the data is used and combined incorrectly, according to these models. Localization and tracking nearby do not frequently result in privacy violations because anyone in the immediate vicinity can legitimately view the subject's region. In general, localization and tracking become a concern during data processing, when area traces are processed at a backend that is not under the subject's control. The primary difficulties in this method are (i) expertise with tracking, including latent data collection, (ii) information control in a shared space in interior settings, and (iii) IoT platforms with privacy-preserving policies(Amani Abuladel. et al.2020).

APPLICATIONS AND SECURITY ISSUES IN WIRELESS IOT DEVICES: WIRELESS SENSORS

The benefits of IoT, such as detecting, processing, communicating, and actuating capabilities, have led to its broad adoption concerns about portability, compatibility, safety and confidentiality, data storage, and device management are on the rise as a result of the widespread use of sensors from many manufacturers, processing platforms, and a variety of communication standards and protocols. To accomplish IoT features such as detection and actuation, distinct identification, accessibility, interaction, cognition, self-configuration and modification, and successful service delivery, the following must be considered (Singh, Singh, and Saxena 2020).

Smart City

IoT sensors are the building blocks of the smart city development like water management, fire alarm, waste management sensors, traffic management, etc. are the wireless sensors that should be secured from malicious attackers. In the smart city concept, IoT devices play a very important role in the traffic management system. These devices manage the vehicle's traffic without any human being's help. This will be a great advantage but it has disadvantages also like whenever a hacker hacks the traffic management system then the person can get all the details of the vehicles which are traveling through a particular location and they can monitor particular person vehicle location all over the city. In this way, the smart city IoT devices may get compromised easily. In a smart city, surveillance plays an important role in lowering crime rates. Surveillance cameras in a smart city record people's activities in congested locations where public safety officials are unable to identify and prevent crimes. This improves public safety. Smart e-governance is also part of the smart city in that it facilitates speedier decision-making in government organizations and initiatives, as well as openness in government agencies and public service accessibility(A. Hameed.2019).

Health Care

The human health status is monitored continuously using many tiny devices which are set up in the human body. These small devices continuously collect the health issues and transfer them to the health care centers. The user shares his medical health-related data and location to the IoT devices and that data going to be stored in the server database. Any hacker who hacks the user ID the hacker can get all the data that they share with the server database. This may cause a threat to the user by the hacker using private data and that may be used by the hacker against the user-health issue by tampering with the health-related data (Axelrod, 2015).

Smart Home

The smart home is achieved only with the help of IoT devices in which sensors like automatic door lock sensors, smart safe management, controlling electronic devices, fans, refrigerators, cooler, air conditioners, washing machines, etc. were used. The small sensors which are used in smart home automation techniques are equipped with very few security features so these small devices can be easily hacked by other external people and they can get the whole security system because those small devices are connected to the central servers (Axelrod, 2015).

Vehicle Transportation

In this generation driverless vehicles were manufactured which are controlled by the user with the help of a remote sensor. Security is a very important thing in this system because a person may hack the vehicle sensors and may easily get control of the vehicle and steal it easily. The private data associated with the vehicle get compromised and that vehicle gets used by the hacker for illegal things so this security is also compromised (Axelrod, 2015).

Fitness Devices

In the current digital age, the majority of users are working for a company and industry that engaged in a range of activities, and as a result, they have a tendency to neglect their health and physical fitness. For patients and athletes, staying fit and eating a healthy diet are critical. Smartwatches, phones, gloves, and smart shoes track features such as heart rate, calories burned, sleep cycle analysis, motion, and so on. IoT allows fitness trainers and instructors to guide athletes and patients based on collected data, resulting in better patient and athlete fitness (Singh, Singh, and Saxena 2020). These devices collect all the health-related personal data about the user and this data can be traced easily by hackers.

Natural Disaster Management

Hurricanes, floods, earthquakes, and other natural disasters are becoming increasingly common, resulting in the loss of habitats, lives, and property. Early warning systems, fast reactions to victims, emergency medical care, and other services are all made possible by IoT technology (Singh, Singh, and Saxena 2020). When assessing environmental parameters, sampling, analysis, geographical, and temporal mistakes are more prevalent. When these errors are removed from the dataset, decision-making becomes more efficient. These sensors generate a massive amount of data that is both heterogeneous and unstructured. As a result, clever algorithms are needed to clean and pre-process this type of data, enhancing analytical capabilities. Implementing lightweight algorithms for big data analytics in environmental monitoring will require more effort. Because of the meteorological features, the lifespan of sensors employed in environmental monitoring is decreasing. Efforts should be made to improve the sensor's durability. Security considerations must be considered in addition to the aforementioned challenges (Phommasan et al., 2019).

SECURITY KEY ELEMENTS OF THE WIRELESS IOT DEVICES: CHALLENGES

Authentication

After gathering the information with the help of sensors the next step is to share this collected data with valid or authenticated devices. Authentication is provided by the server database manager who will be going to receive and use the database. The hackers can steal the authorized person's data and can access the server database using identity theft (Hu et al., 2011).

Access Control

Accessing the resources from the server database is limited to only authorized users. More devices are connected to the network and more data has been generated and collected using these devices and collected data should be accessible by the authorized user only (Schurgot. et al., 2015).

Data Security

The data sharing from one device to another must be through a secured path with encryption technology so that no third person can access the data. The data should not be manipulated by external factors (Solangi et al., 2018).

Integrity

The data has been collected using sensors and that collected data should be shared without being tampered with or changed while moving from sending to receiving devices.

Non-Repudiation

The sender and receiver must be honest concerning the data sharing that they should agree that they have sent and received the data from other IoT devices.

Reliability

Even in the face of failures, an IoT middleware should remain active. To achieve a high-level framework with unshakable quality, every part of the middleware must be approved, including correspondence, information, advancements, and gadgets from all tiers(Amani Abuladel. et al.2020).

Elliptic Curve Cryptography (ECC)

ECC is a lightweight and efficient security mechanism for IoT networks (Tank et al., 2016). It is dependent on the elliptic curve's algebraic structure across finite fields. To encrypt and decrypt communications sent across IoT devices, it employs a method to produce public and private keys, in this case, the transmitter station encrypts data with the recipient's public key, and the recipient decodes it with its private key. The main advantage over the ECC is that it creates a shorter key size as compared to other algorithms so it increases the computation speed and reduces the required time, and memory space (Tank et al., 2016).

Security and Privacy

The capacity of a system to adjust to hostile external or internal assaults, by implementing firewalls, creating processes for inspection and approvals, and employing cryptography can help to increase security. Several devices are communicating with one another. They exchange a large deal of information. This information might be personal and private, especially in everyday life. As a result, Early-stage issues such as confidentiality, safety, and protective measures should be considered in IoT middleware configurations (Abuladel et al., 2020).

KEY TECHNOLOGIES

Blockchain

The blockchain ensures the integrity and data security in IoT devices. The blockchain maintains the distributed ledger in which it records all the data which flows from one device to another. In this system, the data or information is divided into small ones and distributed over multiple computers so data security has been achieved. This technique is also called a decentralized database system. This technique increases the redundancy by storing multiple copies on multiple systems so that whenever any system gets crashed and lost data that data could be recovered using another system. This technique also applied load balancing so that no one IoT device gets dependent on a single system and the workload of the devices gets distributed over multiple computers. This will increase the work speed and decreases the traffic on the particular nodes. Even blockchain also has some disadvantages over the network like lack of trust in data storage, the complication in development, security, and privacy concerns.

Cyber Security

Security is the foremost thing in IoT devices. The technical era produces many useful IoT devices which are useful in day-to-day life but with little or no security features. Because of this reason many IoT devices are vulnerable to their privacy and these devices are easily hacked, and controlled by other unauthorized users which is very destructive and harmful.

RESEARCH OPPORTUNITIES

Artificial Intelligence and the Internet of Things

Several companies have already used AI and IoT in their operations and products. According to a recent Technical Trends study taken by SADA Technology, the Internet of things and Artificial things are the most widely used technologies. It was also discovered that AI and IoT are the leading innovations that businesses are investing in to boost efficiency and gain a competitive edge. IoT is primarily concerned with sensors placed in machines that transmit large volumes of data over the internet. The five basic processes that each IoT-related application must follow are: develop, share, combine, assess, and respond. Without suspicion, the "Action's" worth is decided by the final analysis. As a result, throughout the analysis phase, a better classification of IoT is formed. Artificial intelligence plays a crucial part in this situation.

Voice User Interface

This domain is the trending technology in IoT devices. Controlling and monitoring the internet of things i.e. the things which are connected with the help of the internet with voice command only.

Minimizing the Size of IoT Devices

The size of IoT devices decreases as technology in this field increases. The growing technology manufacturing the things smaller than usual so that those devices get fit in any space and making them reliable in such a way that they should survive and provide better results in any situation. The size of the IoT device should be less and performance must be more.

Power Production and Consumption

The IoT devices are worked or performed with the help of energy which it provides. The small or big device should consume the least energy possible. The IoT devices are made for human convenience and if these devices consume more energy then their valuable data should not become the human profitable thing. So to achieve this goal, the energy needed by the IoT devices must generate themselves.

Big Data and the Internet of Things

In IoT connected with millions of things or devices each other and with a central server. The technology is growing with the help of fast internet and fast internet supports the smart devices which are connected and controlled remotely with the help of the internet. To reduce the work performed by human beings IoT devices are used in day-to-day life. These devices can sense, act, and share the data with a central database for future use. These stored data lead to big data analysis. Big data analysis is used to analyze the stored data and used in future predictions and for many more things. The security of stored data is also important to maintain the private data of users.

CONCLUSION

This chapter reviews the privacy and security aspects of wireless IoT devices, it includes an introduction to IoT and wireless devices, existing research on wireless devices, layers of IoT devices, architecture of wireless devices, various attacks on wireless devices, storage systems, communication standards, challenges, key technology, and research opportunities.

ACKNOWLEDGMENT

Mr. Manojkumar T. Kamble is thankful to the Department of science and technology of Karnataka(DST) for supporting our work through Ph.D. fellowship No. DST/ KSTePS/Ph.D. Fellowship/PHY-04:2021-22/1033.

REFERENCES

Abuladel, A., & Bamasag, O. (2020, March). Data and location privacy issues in IoT applications. In *2020 3rd International Conference on Computer Applications & Information Security (ICCAIS)* (pp. 1-6). IEEE.

Alamer, A. (2021). Security and privacy-awareness in a software-defined Fog computing network for the Internet of Things. *Optical Switching and Networking*, *41*, 100616.

Anajemba, J. H., Yue, T., Iwendi, C., Chatterjee, P., Ngabo, D., & Alnumay, W. S. (2021). *A secure multi-user privacy technique for wireless IoT networks using stochastic privacy optimization*. IEEE Internet of Things Journal.

Axelrod, C. W. (2015, May). Enforcing security, safety and privacy for the Internet of Things. In 2015 Long Island Systems, Applications and Technology (pp. 1-6). IEEE.

Github. (2022). *WLAN Overview (Roaming- current and future enhancements)*. http:// what-when-how.com/roaming-in-wireless-networks/wlan-overview-roamingcurrent-and-future-enhancements/

Goyat, R., Kumar, G., Saha, R., Conti, M., Rai, M. K., Thomas, R., & Hoon-Kim, T. (2020). *Blockchain-based data storage with privacy and authentication in internet-of-things*. IEEE Internet of Things Journal.

Gupta, S. S., Khan, M. S., & Sethi, T. (2019, June). Latest trends in security, privacy and trust in IOT. In *2019 3rd International conference on Electronics, Communication and Aerospace Technology (ICECA)* (pp. 382-385). IEEE.

Hameed, A., & Alomary, A. (2019, September). Security issues in IoT: a survey. In *2019 International conference on innovation and intelligence for informatics, computing, and technologies (3ICT)* (pp. 1-5). IEEE.

Hu, C., Zhang, J., & Wen, Q. (2011, October). An identity-based personal location system with protected privacy in IoT. In *2011 4th IEEE International Conference on Broadband Network and Multimedia Technology* (pp. 192-195). IEEE.

Mohanta, B. K., Jena, D., Ramasubbareddy, S., Daneshmand, M., & Gandomi, A. H. (2020). Addressing security and privacy issues of IoT using blockchain technology. *IEEE Internet of Things Journal, 8*(2), 881–888. doi:10.1109/JIOT.2020.3008906

Phommasan, B., Jiang, Z., & Zhou, T. (2019, September). Research on Internet of Things Privacy Security and Coping Strategies. In *2019 International Conference on Virtual Reality and Intelligent Systems (ICVRIS)* (pp. 465-468). IEEE.

Schurgot, M. R., Shinberg, D. A., & Greenwald, L. G. (2015, June). Experiments with security and privacy in IoT networks. In *2015 IEEE 16th International Symposium on a World of Wireless, Mobile and Multimedia Networks (WoWMoM)* (pp. 1-6). IEEE.

Singh, S., Singh, K., & Saxena, A. (2020). Security Domain, Threats, Privacy issues in the Internet of Things (IoT): A Survey. *Proceedings of the Fourth International Conference on I-SMAC (IoT in Social, Mobile, Analytics, and Cloud) (I-SMAC),* 287–294.

Solangi, Z. A., Solangi, Y. A., & Chandio, S., bin Hamzah, M. S., & Shah, A. (2018, May). The future of data privacy and security concerns in Internet of Things. In *2018 IEEE International Conference on Innovative Research and Development (ICIRD)* (pp. 1-4). IEEE.

Swamy, S. N., & Kota, S. R. (2020). An empirical study on system level aspects of Internet of Things (IoT). *IEEE Access: Practical Innovations, Open Solutions, 8,* 188082–188134. doi:10.1109/ACCESS.2020.3029847

Tank, B., Upadhyay, H., & Patel, H. (2016, August). Mitigation of privacy issues in IoT by modifying CoAP. In *2016 International Conference on Inventive Computation Technologies (ICICT)* (Vol. 3, pp. 1-4). IEEE.

Vakhter, V., Soysal, B., Schaumont, P., & Guler, U. (2022). Threat Modeling and Risk Analysis for Miniaturized Wireless Biomedical Devices. *IEEE Internet of Things Journal, 1.* doi:10.1109/JIOT.2022.3144130

KEY TERMS AND DEFINITIONS

Artificial Intelligence (AI): AI refers to making the machine think and behave like a human being.

Cyber Security: Protecting and preventing unauthorized access to electronic devices, and data in the network.

Denial of Service (DOS): In this method, the attacker sends unnecessary garbage requests to the server and the server gets down because of garbage requests. When an authorized user wants to access and use the server using the internet of things then the server denies the user's request. This process is called denial of service.

Distributed Denial of Service (DDOS): DDOS is the extension of DOS. In DDOS the attacker sends the unnecessary garbage request to the server with multiple IoT devices so that the server gets down faster than the DOS technique.

Internet of Things (IoT): The IoT is a collection of electronic devices which are connected, controlled, and monitored using another electronic device with the help of the internet. These devices exchange data between each other and between server databases.

Chapter 7
Application of Blockchain Technology in an IoT–Integrated Framework

Lipsa Das
Amity University, India

Smita Sharma
 https://orcid.org/0000-0003-0067-9853
Amity University, India

Suman Avdhesh Yadav
Amity University, India

Khushi Dadhich
Amity University, India

ABSTRACT

The IoT is a current technology that has the ability to interconnect the embedded devices and merge the sensors and wireless networks to communicate and exchange data with one another via the internet. Building and automation, healthcare, monitoring, agriculture, etc. are applications of IoT. Currently, with the rapid increase of IoT applications, we created an environment for device-to-device interconnections, new business models, where these interconnected devices accumulate, process, and share the data with each other. Data integrity or ownership issues, cyber-attacks, single point of failure, etc. are some of the limitations of current IoT solutions. Blockchain technology or a distributed ledger technology has the potential to greatly enhance the security aspect, privacy, and reliability concern of the data. High efficiency, transparency, low cost, and no third-party interventions make this technology a one-stop solution for limitations to integrated IoT devices.

DOI: 10.4018/978-1-6684-3921-0.ch007

INTRODUCTION

History of IoT

The IoT (Internet of thing) refers to the network of billions of physical devices those have the internet connection. These physical devices or group of objects are implanted with sensors, electronics, software, processing ability and other technologies which permit them to collect data and share it with other devices or systems which are connected to the internet without any human interaction. Basically we can say that IoT is the interconnection of devices over the internet which can be used to implement the functionality in the objects which we use in our everyday lives by enabling them to communicate with each other by sending or receiving data. Nearly any physical object with internet connection to share information can be converted into an IoT device (Steve Ranger, 2020).

The use of IoT devices on a large scale assure us to convert many aspects of the way we live. Nowadays a light-bulb which can be switched on by using our smartphones. An IoT device can be a small thing such as a small toy or it can be as large as an automatic vehicle. The word IoT is generally used for the devices which are connected over the internet and able to communicate with each other independently within the network without any human interaction. For this cause, neither a PC nor a smartphone is considered as a IoT device. For the consumers, new IoT devices such as: home appliances with internet-connection, home automation components, etc are taking us towards the vision of "Smart Home concept", providing with more security as well as efficiency (Wikipedia). The IoT devices that are made for personal use such as wearable fitness bands and the applications that helps to monitor the health, are making a remarkable difference in the field of health care services. This technology is a great help to the elder people and people with disabilities by providing them independent and enhancing the life quality at a reasonable rate. IoT systems such as traffic systems, networked vehicles and sensor implanted on roads bring us more closer to the idea of "Smart Cities". IoT technology provide the ability to minimize congestion and energy consumption. Many companies and organizations has provided a broad data about the possible influence of IoT on the economy & internet in the upcoming years. For example- Huawei predicts hundred billion devices with internet-connection will be there by 2025 whereas, McKinsey Global Institute predicts that the financial effect due to IOT on the overall global economy may be as much as $3.9 to $11.1 trillion by 2025. All these predictions collectively provide a idea of growth and effect of Internet of things in the upcoming years (Karen Rose et al., 2015)

The intension of attaching sensors & intelligence to the physical devices that was talk throughout the 1980's & 1990's. Due to not having a very robust technology at that time, many project's progress was very slow. IC's and devices are big & bulk, which are not very effective for communication purpose. At early days this idea was termed as "Embedded Internet" or "Pervasive Computing". In year 1999, the term "Internet of Things" was discovered by "Kevin Ashton" during his work at "Procter & Gamble" & to have the attention of his senior management towards the new technology known as RFID while he was working in supply chain optimization. As the internet was the topic of conversation or headlines during that time he prepared a presentation on it and called his presentation as "Internet of Things". Although Kevin grabbed the attention of some of his P&G executives, but still it took at least another decade for the technology to get widespread attention (Knud Lasse Lueth, 2014). "Human culture or our things and digital information system i.e., the internet are now very well integrated by IoT". This is what Ashton told about Internet of things on ZDNet. Adding the RFID tags to the to the equipment's which are expensive to help to track their location was one of the first IOT application. But after that the cost of adding sensor to any object or connecting it to the internet has continued to fall and experts predict that this functionality could one day cost as little as 10 cents, by making it possible to connect nearly everything to the internet. The Internet of Things was initially very beneficial to manufacturing & business, and its application is known as M2M (machine-to-machine). But now it is more used in equipped our home space and office space with smart appliances, by transforming it into something which is relevant to almost everyone (Steve Ranger, 2020).

ESSENTIAL CHARACTERISTICS OF IOT

Connectivity

The term connectivity doesn't need much explanation as the term itself explain everything. Connectivity plays an significant role in the infrastructure of IoT. The things of IoT must be connected to its infrastructure. With everything going on in IoT devices with sensors and connected to a control system, there must be a connection between each and every level. The connectivity of IoT devices should be guaranteed all the time as it can be required at anytime, anywhere and by anyone. Without connectivity nothing makes sense in IoT.

Intelligence

The extraction of knowledge from the data which is collected by IoT devices is very important. Let's take an example, each sensor in the different IoT devices produce or collect data, but that data will only be useful when it is interpreted properly.

Identity

Each IoT devices has it's own identity which makes it different and unique from other devices. This identity is helpful in tracking the device and also at the time of checking it's status.

Complexity

The IoT is considered and studied as a complex system. IoT devices should adapt themselves dynamically according to the changing environment and structure. Talking about a practical approach, all the IoT devices are not connected at the global environment. The Subsystems are always executed to reduce the risk of control, reliability & privacy e.g., the robots which are used at domestic level need to share data within the organization and need to be available within a local network (wikipedia).

Scalability

The number of things connected over the internet is increasing day by day. Hence, the IoT setup should be able to handle this large growth at the same time. The data generated by the IoT devices is huge and it should be handled properly.

Architecture

The architecture of the IoT should not be homogenous in nature. It should be compound so that it can convert itself accordingly to the different function in the IoT network. The IoT is not owned by any single engineering branch, it is a platform where multiple estates come together.

Safety

When a person's device is connected over internet then, there is always a risk of loss of sensitive personal details of the user which can harm the user to a great extent, so the safety of data is a major concern in the field of IoT .

TECHNOLOGIES INVOLVED IN IOT

As we have discussed earlier, IoT is a system of device which are interconnected with each other or objects which are provided with UID and the ability to collect and share information with other devices have internet connection, without any human interaction. The aim of IoT is to create a smart environment where each and every individual can live in a smarter and in a comfortable way. IoT minimize the distance between the physical and virtual worlds & become a important part of our daily lives, so to understand how it works we can divide the IoT technology stack into several layers which are involved in making the IoT work.

Device Hardware

In IoT the term things represent the devices. Playing as a junction between the digital world & real world, can be of different shapes, sizes and levels of difficulties according to the assignment they are asked to execute. Regardless of a small size earphone or large machines every object can be converted into connected devices by adding sensors or actuators to it so that it can share or collect the necessary data.

Device Software

It is the software that makes the IoT device smart. Software is accountable for executing the communication with cloud, gathering information and also do implementing the analysis of real time data within an IoT network. It is also responsible for the users to visualize information and interaction with the IoT network.

Communication

After the proper arrangement of the hardware & software in a device, it requires one more layer which is responsible for the connected devices to exchange information with the other IoT devices. Communication layer include the physical connectivity solutions as well as protocols required in IoT systems. Choosing the right communication solution is an essential part in IoT technology. The selected technology will decide the method for receiving the data or shared from the cloud, also it will make the decision how the IoT device will communicate with the third party devices. Based on each IoT application, different communication option provide various service scheme depending upon the power consumption, range and bandwidth.

CONNECTIVITY SOLUTIONS WITHIN THE IOT TECHNOLOGY

Short Range IoT Network Solutions

Bluetooth

For establishing the short range connectivity, Bluetooth can be consider as the most appropriate solution for connectivity, as on the basis of the future of the wearable electronics market such as headphones, health bands and smartphones. Keeping low cost and low power consumption in mind Bluetooth is designed with Bluetooth Low Energy (BLE) protocol which require very less power from the device. It also has a disadvantage at the same time, while transferring a large amount of data this may not be a most successful solution.

RFID

RFID stands for Radio Frequency Identification. It is the first ever IoT application which is used. In logistics & supply chain management, RFID offers the solution for IoT applications, which need the capacity of calculating inside object position any building or organization. The possible application of RFID in real time is tracking a patient in hospital to keep a status about his health and condition.

MEDIUM RANGE IOT NETWORK SOLUTIONS

WIFI

It is the most commonly known wireless communication protocol. It's widely used across the IoT network it is limited by the higher power consumption result in the need of high signal strength and fast data transfer for better connectivity. WIFI provide extensive ground to the number of IoT solutions.

ZIGBEE

ZIGBEE is the most commonly use wireless system. Its application is mostly used in traffic management system, machine industry & household electronics. It is very low at cost and power operation. Also it provides security and reliability.

Thread

It is especially designed in home products so that that devices that are connected are able to interact with each other, access services in cloud and also communicate with the user through mobile application.

Long Range Wide Area Network (WAN) Solutions

LoRaWAN

LoRaWAN stands for Long Range Wide Area Networking Solutions. It consumes low power to share data with large number of devices over a large network. It is very low at cost and provide bi-directional or both way communication system within industrial applications as well as smart city.

NB-IoT

It is a new radio technology that guarantee consumption of low power and give connectivity with signal strength approx. 23dB. The network infrastructure existing before used by this to provide not only very wide range of coverage but also best quality of signal for the fastest data transmission.

Platform

It is the place where all information is collected, processed, analysed and represented in an accessible manner. The information collection or processing not only the things that makes it unique but its ability to find useful perception from the information which is provided by the IoT devices by the communication layer. There are many IoT platforms available in the market, the choice of the IoT platform depends upon the IoT project, architecture, technology, protocol, security and hardware and software use in different IoT devices. Platform can be either installed or it can be cloud based (Sunil, et al, 2020).

Internet of Things (IoT) enabling technologies are:

1. **Wireless Sensor Network (WSN):** A WSN includes the devices which have sensors which enable us to check the physical and environmental condition. It basically comprise of end nodes, routers and coordinators. End nodes consists of seveal sensors where the information is transferred to coordinator with the help of router. Example of WSN are surveillance system, health monitoring system, etc.

2. **Cloud Computing:** Cloud computing gives us the access to the applications according to its use over the internet. The word cloud states something which is in remote location. With the help of cloud computing, users can acquire sources from anywhere on the internet such as database, web servers, etc. Cloud computing has broad network access and measured services (Alam et. al, 2020).

   Different services provided by cloud computing are:
   - **IaaS (Infrastructure as a service):** IaaS supply online services such as virtual machines, networking and data center space on pay per use basis. Example of IaaS providers are Google Compute Engine, Amazon Web Services, etc.
   - **PaaS (Platform as a Service):** PaaS gives a cloud-based environment to give support to the cloud applications without giving any price for buying and managing hardware and software. Platform such as hardware, operating system, etc provide a platform to develop these applications. Example are app cloud, google app engine.
   - **SaaS (Software as a Service):** SaaS provide a route to deliver application on the internet as a service. Instead of installing software, we can directly use them through the internet. SaaS applications are also called web based software or demand software or hosted software. Example are Google Docs, Gmail, etc.

3. **Big Data Analytics:** Big Data Analytics is a process to study volume of a data or big data. Gathering of data whose volume is too huge that it become difficult to store, process and examine that data using traditional database. Through big data it become easier to handle huge amount of data. Example bank transaction, E-commerce, etc.

4. **Communication Protocol:** Communication protocol help connected devices to share and collect data over the internet. Multiple protocols are used for different single communication. Protocol suite is designed in such a way that different protocols can work together, when it is applied on a software it is called protocol stack. It is highly used in data encoding and addressing schemes.

5. **Embedded Systems:** Embedded system is a union of hardware and software which are used together to perform specific task. It consist of microprocessor memory, networking unit and input and output devices and storage unit. It gathers the data and share it over the internet. Example are digital camera, wireless router, etc.

Figure 1. IoT system architecture (www.digi.com)

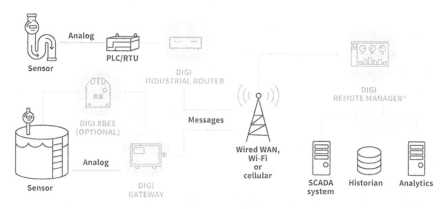

IOT SYSTEM ARCHITECTURE

The architecture of IOT system is basically divided into a 4- stage process in where the data flow from the sensor which is attached to the objects or things and then finally this data is provided to data center or to the cloud for process, storage & analysis purpose (Vikash Kumar, et al, 2021).

Stage 1: Sensors and Actuators

Here the process initiates with actuators & sensors, the connected devices that controls the physical process in the IoT devices. These sensors collects the information regarding the status of any physical process or environmental condition, such as humidity or temperature. In some cases sensors discover a situation where there is requirement of immediate reaction. For example, robots working in industrial area need to give certain actions to a situation in real time. In these situations, there is a requirement of very low latency between the sensors and actuators, to minimize the data transfer delay, so that the data analysis can be done without any failure.

Stage 2: Internet Gateways and Data Acquisition Systems

DAS gathers the raw information from the sensors or actuators. This generates a digital format from the data before transmitting it to internet gateway via wireless WAN or wired WAN for the next stage of processing. At this stage the volume of data is at the maximum level. The data can be large, e.g., in a industry where hundreds of sensors collect information simultaneously. Hence for transmission purpose, the information is need to be compressed to the finest size.

Stage 3: Pre-Processing - Analytics at the Edge

Once the DAS converts the information into digital format, it require processing so that the volume of the data can be reduced further before sending it to the data or information center or cloud. The edge device do some pre-processing at this stage. Machine learning plays a very important role at this stage by providing observation to the system and enhance the process on the basis of the on-going procedure, without waiting for any further instruction from the cloud or the data center. This processing take place on a device to a location which is closer to the sensors, such as on-siting writing closet.

Stage 4: In-Depth Analysis in the Data Center or Cloud

Powerful IT systems participate at this stage so that the data can be analysed, managed and stored securely. This process is executed in the corporate data center or in the cloud, where data from different sensors is combined to provide a large vision for the overall IoT system. Industry specific or company specific applications can be used at this stage to perform in depth analysis. This stage also consist of storage of the data in warehouse for further analysis and record keeping.

IOT DEVICES

As we have discussed above, any physical object that is attached to a sensor and is able to collect or share data with other devices and is able to perform task over the internet without any human interaction is termed as IoT devices. Many IoT devices allow the users to operate them according their needs whereas some IoT devices operate themselves accordingly to situation. There are different type of IoT devices in the market, let's discuss about some commonly used IoT devices.

- **Smart Home Appliances:** These devices are designed according to the consumer needs. These devices are designed in such a way that they can help the user in household work and make their work easier. Example of such devices are speakers, fridges, etc. Smart home device market has reached around $23.328 billion in 2020 and it has been declared as the major contributor in the IOT device market.
- **Industrial Sensors:** These devices are constructed for the industry purpose so that the manufacturers can collect the information about their machines or to monitor problems in a factory by using internet connected sensor

devices. These IoT sensors can improve the maintenance of the machines and operational visibility and the consumption rate for specific resources.

- **Smart Automobiles:** In the effort of creating self-driving vehicles, cars and trucks of all sizes are using the IOT functionality as a major part of it. The investment in autonomous vehicle industry has reached over $100 billion. It is the first company to release a self driving car.
- **Smart Cameras:** Smart cameras are used at both private and business level. These days more internet connected cameras are developed so that the recordings can be stored safely onto the location.
- **Manufacturing Robots:** Manufacturing of robots is increased these days. As these robots has the capability to perform different task in a more accurate way. These robots can be remotely controlled and programmed, according to the need of the manufacturer or the user.
- **Healthcare Devices:** As people are getting more busier day by day that they are not able to take care of their health. IOT technology has developed some devices such as fitness band so that the user can keep a track of their health status. It collects all the different information about the health which makes the work more easier. The data which is collected can help the user to improve their health and live a healthy life.

IOT ADVANTAGES AND APPLICATIONS

Advantages of IoT

- **Monitor Data:** The main advantage of IoT is monitoring. It help us to keep the track on the data which has been collected by the device. For example taking an example of a air purifier at home which is connected to your phone so it helps us to the track of air quality at our home. It collect all the data and store it safely which is not possible by human to do so.
- **Ease of Access:** Today we are able to gain the information about anything at anytime, regardless of where ae are. T just require a smart device and internet connection. Now days we use google maps instead of asking location to a person in real life. Accessing information about recent research, business or anything has become much more easier it takes only few seconds to search about anything and gather information about it.
- **Speedy Operation:** All these IoT devices by collecting all the data help us to complete multiple task with a fast speed. Smart industries use IoT devices to perform the automate repetitive task so that the employees can focus on different or more challenging things.

- **Better Time Management:** We can look up to the news at our phones during our daily lives, we can purchase an item by online shopping instead of going to the shop physically, we can almost do everything from the palm of our hands. IoT has made all these things possible. With the help of IoT we are able to save our time and invest it at other things which helps us to improve our skills.
- **Save Money:** The main reason behind the IoT to be accepted widely is that the cost of monitoring machines is less than the amount of money saved. IoT helps the people in their daily lives by communicating with each other in an efficient way by saving their cost and energy.
- **Automation and Control:** Due to physical objects connected to the internet and with wireless technology there is a huge amount of automation and controlling. But without any human interference devices are able to communicate with each other and able to complete task more efficiently (Soumyalatha et.al, 2016).

Applications of IoT

1. **Consumer Application:** IoT devices which are developed according to the need of the consumer such as home appliances, wearable technology connected vehicles with remote monitoring capabilities (Kiran et. al, 2018).
2. **Smart Home:** The most popular application of IoT is in smart home. Now days people love to connect their home appliances such as lighting, air conditioner, security system, camera system, etc connected to their smartphones. These IoT devices are becoming more and more popular in coming generation as it provide you complete freedom to design you house in a way you want. These devices are becoming so popular that in every second new 127 devices are getting connected to the internet. Some of these popular devices are google home, Philips Hue lighting, etc.
3. **Elder Care:** This application provide a way to take care of your elder ones or the person with disabilities. These applications include voice control which can be helpful to the people who are aged and also to the people who are not able to walk. Sensors such as medical emergency are also very helpful. In this way we can provide user a high quality of life (wikipedia).

Organizational Application

- **Healthcare:** Many IoT applications has been developed in the industry of health care where doctors can check their patients through web cameras or machines without coming in direct contact with them or meeting them personally. During this time of pandemic it is a very useful application as you can talk to the doctor from home only. Robots are highly used in health care. It include surgical robots that can help the doctor to perform surgery more carefully and effectively. Other types of robots also include nursing robots which can keep the track of the patient health and monitor them from time to time.

- **Transportation:** Self driving vehicles are highly dependent on IoT. These vehicles have a lot of features and function that requires sensors to communicate with each other to handle navigation, speed, etc. The technology used in self driving vehicles needs to be very accurate so that each and every device can communicate with each other reducing any chances of accidents. Tesla is the most popular company in terms of self driving cars. These days we can book transportation by sitting at home such as cabs, bus, train tickets, etc. Application such as ola and uber provide us the facility to book cab at anytime and anywhere and we can also communicate with the driver.

Industrial Application

- **Manufacturing:** With the help of IoT intelligence which is used in industry it enable us to manufacture new products in a short period of time and also help to give quick response to the increase in the demand of the product
- **Agriculture:** Food is a very important part of our life without it we cannot live. It is very unfortunate that we waste this food some time. Countries like Sudan, Chad, etc are the poorer countries in terms of food. The one way to check the food availability in different countries can be done with the help of IoT devices. It can be done by collection data about the seed type, soil quality through different sources such as satellites, local weather stations, etc. Than according to this data we can advice different machines and farming techniques. All this will result in better crop production. All this work can be done so easily and accurately with the help of IoT only.

Infrastructure Application

- **Smart Cities:** The save energy and use the resources less, more well organized cities can be constructed with the help of IoT. This can be possible by the combination of different sensors that are used to perform different tasks such as traffic management, waste management, creating smart buildings, etc. Countries like Singapore, Oslo are getting more connected to the IoT and becoming the smartest country.

- **Smart Pollution Control:** Pollution is one of the major and biggest problem in the cities. Sometimes it feel like we are breathing in smog instead of oxygen. This pollution level can be controlled with the help of IoT. First of all we need to collect all the data regarding the pollution like emission from vehicles, airflow direction, traffic, etc using the combination of different sensors. Using this data we can conclude that what we can do to solve this problem (Kiran et. al, 2018).

IOT ISSUES AND CHALLENGES

Internet of Things (IoT) has now become a major part of human lives, communication and business. But at the same time there are different challenges which are faced by IoT (Shevale Rupali et. al. 2020).

Security Challenges in IoT

Lack of Encryption

Encryption is considered as one of the best way to prevent hacker from accessing data, but at the same time it is considered as one of the challenges which is faced by IoT. It is easy for a hacker to change the algorithm which are developed for the protection. As the storage device & processing capabilities can be easily detect on a computer (Abomhara et. al, 2002).

Insufficient Testing and Updating

As the number of IoT devices is increasing day by day, manufactures are more interested in producing more and more IoT devices to full fill the needs of the consumer so that the product can be delivered as soon as possible without providing security facilities with the device or without taking much care about it. As these devices does not get enough testing and update, it become easier for the hacker to hack these devices.

Brute Forcing and the Risk of Default Passwords

Weak authorization and login method make all the IoT devices more prone to password hacking. Any company or organization that use default authorization for all the devices which are at their company, by doing this they are placing business assets and consumer personal details both at risk.

IoT Malware and Ransomware

Ransomware uses encryption method to remove user from various devices and platforms and still use their id as a user and get valuable data and info about the user. For example a hijacker can hijack your camera and get all the pictures which are taken by you. With the help of malware accessing, the hackers can unlock the device and return the data.

IoT Botnet Aiming at Cryptocurrency

The workers which are working at IoT botnet can change the data privacy, which is a very big risk for any open crypto market. The blockchain companies are trying their best to increase the security.

Design Challenge in IoT

Increased Cost and Time to Market

Implanted system are little bit strained by the cost. It is required to provide better approach while designing an IoT device. In order to full fill the needs the cost of modelling and the cost of implanting electronic components in a device should be handled. Designers also need to reduce the time of designing a device so that the device can reach to the market at the right time.

Security of the System

System which are designed and embedded in the device must be secure with cryptographic algorithm and pass through all the security procedures. It consist of various approaches so that all the components which are implanted in a device can be secured from prototype to deployment.

Deployment Challenges in IoT

Connectivity

It is one of the leading concern while connecting devices, applications and the cloud platform. The IoT devices which are connected provide the data which is extremely valuable. If the connectivity of the device is poor it become one of the biggest challenge as the IoT sensors need to monitor process data and share information.

Cross Platform Capability

IoT application must be developed by considering the technological changes of the future. This development need a balance between the hardware and software functions. It is a big challenge for the IoT developer to create a application which is able to give best performance despite heavy device rates and fixings.

Data Collection and Processing

The data plays a very crucial role in IoT application. Along with security, it is important to make sure that the data is collected, processed or stored within the suitable environment.

Lack of Skill Set

The development challenges which are mentioned above can only be controlled if there is appropriate resource which is working on the IoT application. The right method will help to pass the major challenges and it will become an important IoT application development asset.

CLOUD ENABLED IOT

One of the reason behind the success of the Internet of Things is Cloud Computing. Cloud computing allow the user to perform various computing task using the services which is available on internet. The use of IoT in coexistence with cloud technology has made both of them related to each other. This is one of the technology which will provide huge benefits in the future. Because of increase growth of the technology, the issue of storing and processing large amount of data has increased. Because of the combination of IoT and cloud technologies, it will be possible to provide powerful monitoring and processing services. For example, sensor data can be saved using cloud computing so that it can be used later for monitoring and activation using other devices (Oknacki et al., 2013).

Benefits and Functions of IoT Cloud

1. IoT cloud computing has issued many different connectivity options which allow the user to access to a large network. Devices such as mobile, tablets, laptops, etc provide the user to gain access to cloud computing resources.

2.   The only requirement to use cloud computing is network access. It is a web service which does not require any permission or help, so developers can use IoT cloud computing very easily.

3.   According to the need of the user storage space can be expanded, we can edit software settings, which makes it fast and flexible. Because of these characteristics it provide deep computing power and storage.

4.   The number of IoT devices and automation is increasing day by day, it also increase the concern about the security of these devices and users personal details. Cloud computing solution provide authentication and encryption protocols for the safety of the device and the information stored in it.

5.   IoT cloud computing is very convenient as it provide the services in return as much as you pay for it. The increasing network of objects needed to be connected to the internet and exchange data with the components in the network which can be done with the help of cloud computing.

6.   Cloud computing helps to improve the productivity of the daily task in coexistence with the Internet of Things. The work of cloud computing is to provide a route to the data so that it can reach to its destination while the IoT generates a huge amount of data (Atlam et. al, 2017).

BLOCKCHAIN FOR INTEGRATING IOT DEVICES

Business and organization runs on the basis of the information & data. The faster the it receive more true information, the better. For delivering the data Blockchain is perfect because it give immediate and completely transparent information stored on immutable ledger that can be accessed by only few people who are allowed to access it. This blockchain network can track order, payment and account etc. (Kshetri et. al, 2017) Blockchain is basically defined as a system where the information is stored in such a way that it is difficult to change, hack or cheat the system. The blockchain is a truly trustless system which remove the third party interference to process the transaction. The processing of such transaction ab any third party could increase the cost of the services. Whereas in blockchain technology the transaction can be done between the two nodes without relying on any institution such as bank, accountant or a CA. This system is called decentralised system as there is no central point for any failure or hacking.

We have millions of IoT devices everywhere around the world and yet there are many more devices to connect with the IoT environment. Many IoT devices and protocols that are used for communication purpose, are challenging to control or any information exchange. Thus because of it IoT is facing many challenges in the present. Hence to have a decentralized model, the application of blockchain can be proved to be and provide more protection to the system. In Blockchain technology, there are blocks that contain the transaction details where each block has the address of the previous block. Each block in the blockchain is identified by a different hash value which is generated by using hash function. This header field consist the hash value of the previous block. The block basically consist of header, metadata and the list of transaction. The size of the header is fixed as 80 bytes, whereas the size of the transaction is not fixed it depends upon the data which is transmitted. As the value of the previous block is stored in the current block, therefore no changes in the transaction can be done in the previous block. If any change occur in the previous block than it will get authenticated by all the nodes which are included in the process. So we can say that each node in the blockchain has its own copy which is constant across the chain.

As IoT is a system of devices which are implanted with sensors and connected with each other through a network over the internet. IoT devices has unique identification by using Electronic Product Code (EPC), URI, etc. Different manufacturer uses different protocol for identification, communication and device management. Thus cryptographic algorithm are not sufficient to provide security to the IoT environment. As IoT is based on internet, so it become quite easy to inherit the data and it increases the data insecurity. The data is stored and accessed from the cloud services in the IoT system which is more prone to cyber attacks such as tampering, SQL injection, etc. IoT system can be protected by using blockchain technology in it. Based on the access of the blockchain network it can be divided into public, private and hybrid blockchain networks. Based on the complexity, all three network of blockchain will be applied to different application.

Protocols have a very crucial role in the performance of blockchain based IoT application. For the open networks PoW based consensus protocols are found to be the most suitable one for the data to be secured. Whereas for the IoT application with many IoT devices and heterogenous environment Ethereum is found to be the most suitable one.

ARCHITECTURE FOR IOT BLOCKCHAIN PLATFORM

There are various limitation to that of the current IoT solutions. Such as:

- Data integrity issues or ownership issues
- Architectures are highly centralized
- Prone to cyber attack
- Single point of failure

A de-centralized method can a solution to these current drawbacks of IoT solutions (Oscar Novo, 2028). The IoT & blockchain can be integrated together with the help of architecture. The data is firstly received in data logger by the sensors. Than the data logger received encrypted data is send to the local node. The local node is combined with a LPWAN gateway. The peers which are available in the network send the data through the gateway and after that the block-chain nodes are used to store the data. We will now discuss about the process briefly

LPWAN Gateway

LPWAN gateway is a computing device which not only powerful but also act as a full node in the blockchain. Responsibility of the LPWAN gateway is that id provide a way to the data inside a network. It also verify the integrity so that a trust-less IoT infrastructure is achieved (Bardyn et. al, 2016).

Local Acquisition Node

The local acquisition node's responsibility is to collect the data which is sent by the data logger. Then the received data from the data logger is encoded to make sure the confidentiality of the system. Local Acquisition node send the transaction to LPWAN which is the full node to solve PoWs.

End Devices

It is true that the functions performed in blockchain are not possible to be embedded in sensor as they only have the ability of sensing. The data which is received in LPWAN gateway is passed to the blockchain infrastructure. This is the way in which the communication can be done with a node without the need of any storage or computational requirement in the IoT network (Fahad et. al, 2020).

CONCLUSION

This chapter describes the various benefits and application of IoT devices over traditional devices. How the Blockchain technology prove to be a blessing to overcome the limitations of modern IoT devices and its potential applications. Using block-chain to store IoT data would add another layer of security that hackers would need to bypass in order to get access to the network. A great level of encryption is provided by block-chain which makes it virtually impossible to overwrite existing data records.

Figure 2. Proposed IoT-blockchain platform system architecture (Fahad et. al, 2020)

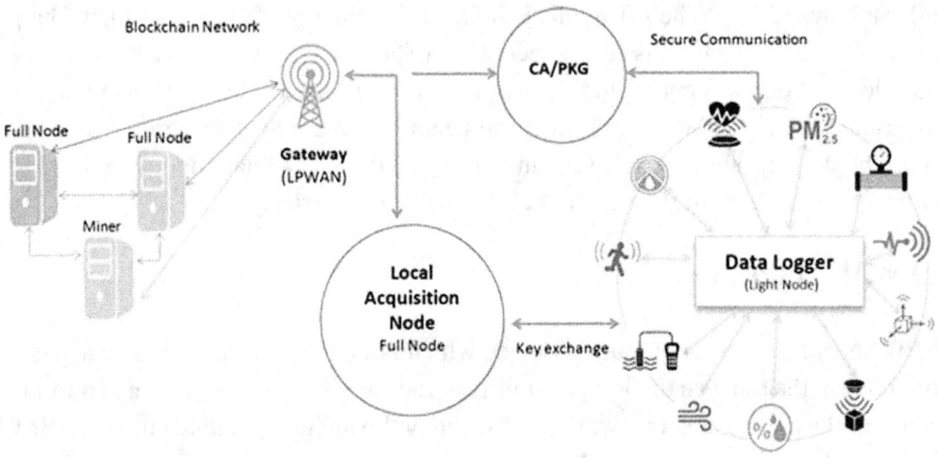

REFERENCES

Abomhara, M., & Koien, G. M. (2014). Security and privacy in the Internet of things: Current status and open issues. In *International Conference on Privacy and Security in Mobile System*. IEEE.

Aggarwal, V. K. (2021). Integration of Blockchain and IoT (B-IoT): Architecture, Solutions, & Future Research Direction. In *IOP Conference Series: Materials Science and Engineering*. IOP Publishing.

Alam, S. (2020). Internet of things (IoT) enabling technologies, requirements, and security challenges. In *Advances in data and information sciences* (pp. 119–126). Springer.

Atlam, H. F., Alenezi, A., Alharthi, A., Walters, R. J., & Wills, G. B. (2017). Integration of cloud computing with Internet of Things: challenges and open issues. *Proceedings IEEE International Conference Internet of Things (iThings) and IEEE Green Computing and Communications (GreenCom) and IEEE Cyber, Physical Social Comput. (CPSCom) and IEEE Smart Data (SmartData)*, 670–675. doi:10.1109/iThings-GreenComCPSComSmartData.2017.105

Bardyn, J., Melly, T., Seller, O., & Sornin, N. (2016). IoT: The era of LPWAN is starting now. In *Proceedings of the 42nd European Solid-State Circuits Conference, ESSCIRC Conference* (pp. 25–30). IEEE.

Booth, G., Soknacki, A., & Somayaji, A. (2013). Cloud security: attacks and current defenses. *Proceedings 8th Annual Symposium on Information Assurance, ASIA13*, 4–5.

Cheruvu, S. (2020). Connectivity technologies for IoT. In *Demystifying internet of things security* (pp. 347–411). Apress.

Khan, M., Quasim, M., Algarni, F., & Alharthi, A. (2020). *Decentralised Internet of Things: A Blockchain Perspective*. doi:10.1007/978-3-030-38677-1

Kiran, S., & Sriramoju, S. B. (2018). A study on the applications of IOT. *Indian Journal of Public Health Research & Development*, *9*(11), 1173. https://doi.org/10.5958/0976-5506.2018.01616.9

Kshetri, N. (2017). Can blockchain strengthen the Internet of Things? *IT Professional*, *19*(4), 68–72.

Novo, O. (2018). Blockchain Meets IoT. *An Architecture for Scalable Access Management in IoT*, *5*, 1184–1195. doi:10.1109/JIOT.2018.2812239

Rupali, Shinde, & Yogita. (2020). Internet of Things Security Risk and Challenges. *IRJET*.

Soumyalatha. (2016). Study of IoT: Understanding IoT architecture, applications, issues and challenges. *International Journal of Advanced Networking & Applications*, *478*.

Chapter 8
Cryptography and Blockchain Solutions for Security Protection of Internet of Things Applications

Kamalendu Pal

(iD) https://orcid.org/0000-0001-7158-6481
City, University of London, UK

ABSTRACT

In recent decades, the industrial applications of the internet of things (IoT) have been attracting massive motivation for research and improvement of industrial operations. The IoT technology integrates various smart objects (or things) to form a network, share data among the connected objects, store data, and process data to support business applications. It is challenging to find a univocal architecture as a reference for different business applications, which can relate to many sensors, intelligence devices, networks, and protocols for operations. Moreover, some of the IoT infrastructural components are a shortage of computational processing power, locally saving ability, and data communication capacity, and these components are very vulnerable to privacy and security attacks. This chapter presents an overview of different IoT-based architectures and security-related issues. Finally, the chapter discusses the challenges of cryptography and blockchain-based solutions after reviewing the threats of IoT-based industry-specific business cases.

DOI: 10.4018/978-1-6684-3921-0.ch008

INTRODUCTION

The innovation of digital communication technology is ushering in a new dawn with the evolution of intelligent electronic devices and high-speed data transportation infrastructures (Pal, 2021a). At the same time, IoT technologies add an extra layer of added advantages to industrial applications. For example, the basic idea of IoT was conceived at the Massachusetts Institute of Technology (MIT), USA, in the late 1990s. The IoT idea consists of self-controlled smarter objects ("things") and their intercommunicability, creating a unique environment and interaction capabilities. For example, intelligent IoT devices can be wearable wristwatches, automobile cars, intelligent factories, radio frequency identification (RFID) tags (Pal, 2019) (Pal, 2021b), and communication technologies are part of this envisioned new world. In addition, there are some prominent IoT-based applications are as follows: (i) smart home applications that include control of domestic electronic objects (e.g., television, refrigerator, security alarm, lighting, heating) and monitoring remotely, (ii) intelligent transport system, which relies heavily on real-time data gathering, and analysis to make judicious decisions for daily transport management and provide appropriate services to the customers, and (iii) digital healthcare services to improve patient-caring capabilities for medical caring agencies.

This innovative progress depends heavily on radio frequency spectrum allocation, other infrastructural issues, and global policymakers' support and wiliness to the technology ratification process. In addition, IoT-based industrial information systems are gaining massive popularity, and these systems use low-power lossy networks, with some devices having constrained resources. Besides, the telecommunication industry is preparing to evolve *wireless networks* to the next generation of technology, known as sixth-generation network connectivity (6G). The 6G wireless networks represent the most meaningful change in the commercial world has seen in wireless networks since *cellular data communication* came into industrial use.

The large scale IoT-based information systems deployment results in significant security and privacy-related issues (e.g., authorization, access control, information security management) (Pal, 2021a) (Pal, 2021b). For example, IoT-based smartphone appliances and other intelligent embedded equipment need to create a digital world for distinct end-user categories for required services. However, privacy and security are not assured in this new physical to the digital world mapping environment. In this way, user privacy maybe not be secured, and often data might be leaked when the data communication channel is interrupted, or data is stolen by the hackers (Pal, 2021a). Consequently, developing IoT-based applications requires carefully addressing various security management solutions (Pal, 2021b).

In recent decades, there has been massive attention to finding appropriate security solutions in IoT industrial applications. Many of these methods aim to mitigate security-related issues at a specific layer of the data communication stack, whereas others aim to enhance IoT infrastructure security at all levels. For example, Fadele Ayotunde Alaba and fellow researchers presented some of these approaches to secure IoT systems (Alaba et al., 2017). In addition, it includes application architecture, communication, and data related issues. Besides, it includes IoT security taxonomy for hardware, network, and application layers.

In the same way, another IoT application review by Granjal and others (Granjal et al., 2015) presented safeguarding problems for the protocols deployed in IoT systems. The security-related issues reviews described in (Roman et al., 2011) (Granjal et al., 2008) (Cirani & Ferrari, 2013) present various necessary operation related management systems and modern cryptography applications. Moreover, other groups of academic authors (Butun et al., 2014) (Abduvaliyev & Pathan, 2013) (Mitchell & Chen, 2014) aim for a comparative analysis of different invasion tracking systems; and in recent years, a particular category of computing (known as *fog computing*) based IoT applications security issues (Yi et al., 2015) (Wang et al., 2015) are also domaining.

A group of researchers (Sicari et al., 2014) highlights the meaningful interpretation of confidentiality, security, IoT applications access control, and middleware-related privacy and security-related issues. The researchers highlight authentication, instruction detection mechanisms, and other relevant security-related aspects of an information system. Different aspects of service-oriented computing (including edge computing-based paradigms) are reviewed by a research group (Roman & Lopez, 2016).

At a conference, Vladimir Oleshchuk discussed privacy and security related issues and presented a survey for IoT-based applications (Oleshchuk, 2009). The same researcher stressed that protecting multi-client computations requires to be enforced to maintain the privacy and security aspects of IoT users. The special techniques (e.g., attribute-based access control) are presented as practical solutions for IoT applications' privacy. Zhou and fellow researchers (Zhou et al., 2017) presented another survey on IoT applications security issues concerning exclusive identification of objects, other privacy-related issues, the requirement for lightweight cryptography techniques, and many other weaknesses. The IoT-a research project (IoT-a, 2016) presents an IoT-based application architecture dealing with trust, security, and privacy deployment. In addition, to avert inappropriate management of data, the privacy model needs to use appropriate protection mechanisms (e.g., cryptography, monitoring access control).

The chapter's main objective is to briefly discuss IoT-related information communication architectures and security challenges for industrial applications. In addition, it also discusses the safety measures regarding data and information sharing in IoT-based systems. Besides, the chapter discusses privacy and security related

Figure 1. IoT connecting technologies and disparate industries

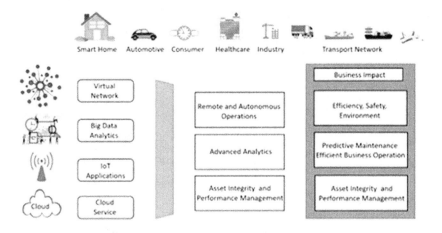

issues using cryptography as a solution mechanism. The chapter also examines the uses of cryptography techniques for resource constraints devices. Finally, the chapter describes a comparative analysis of different lightweight encryption techniques.

IoT ARCHITECTURES AND SECURITY CHALLENGES

The growth of IoT-based industrial applications is steered by business requirements as part of industrial digital transformation, as shown in Figure 1. An ideal IoT-based application comprises devices with embedded sensors and communication connectivity. As IoT uses an extensive range of technologies, it is impossible to prescribe a single reference architecture that can be used as a standard for all conceivable implementations. Therefore, many reference architectures are available for IoT-based industrial application deployment purposes.

Depending on the scale and cross-industry use of IoT-based information systems, data creates different types of enterprise architectures. For example, a simple categories IoT-based information system is shown in Figure 1.

A typical IoT deployment contains heterogeneous devices with embedded sensors interconnected through a network. The devices in IoT are uniquely identifiable and are mostly characterized by low power, small memory, and limited processing capability. The gateways are deployed to connect IoT devices to the outside world for the remote provision of data and services to IoT users.

Figure 2. Different layered architectures

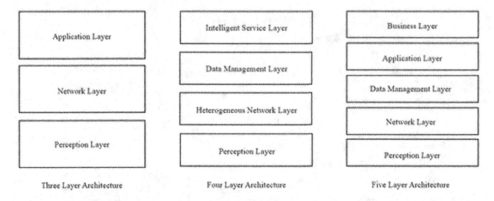

Evolution of Enterprise Information Systems Infrastructure

The primary functions of a simple business information system are (i) to collect dynamic and static data from items of business interest, (ii) to analyze data using computational models; and (iii) to create the strategy and manage a system and optimize system performances using the processed data. A classification of IoT components is presented by NIST (National Institute of Standard and Technology, USA). Moreover, there is no unique agreement on architecture for IoT-based applications. Many researchers have presented various architectures. This section describes a few IoT-based computational architectures, as shown in Figure 2 and Figure 3.

Three-Layer Architecture

Figure 2 presents a three-layer architecture proposed by Eleonora Borgia (Borgia, 2014), which discusses the basic functionalities of an IoT-based system. This architecture comprises the application layer, network layer, and perception layer. The perception layer relates to the collection of operational data, the network layer is responsible for data communication purposes, and the application layer has decades of industry-specific applications. A group of researchers (Said & Masud, 2013) presents a different layered architecture encompassing the perception, heterogeneous network, data management, and intelligent service layers, as shown in Figure 2.

Further on, practitioners and academics explore a five-layer IoT application architecture. Figure 2 diagrammatically represents this architecture, consisting of a perception layer, network layer, data management layer, application layer, and

Figure 3. Fog architecture of an innovative IoT gateway

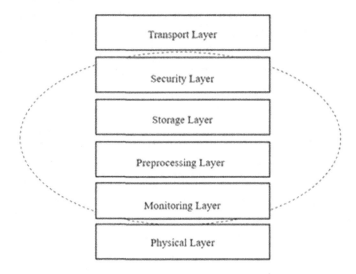

business layer (Aazam & Huh, 2014) (Stojmenovic & Wen, 2014) (Bonomi et al., 2012) (Bonomi et al., 2014).

In recent decades, a group of researchers produced a particular architecture that uses the concepts of service-oriented computing and edge computing related ideas (Gubbi et al., 2013) (Pal & Yasar, 2020b). This architecture is a bit different from its predecessors, and it can provide smart application services using an IoT-based information system. Figure 3 depicts this innovative architecture.

This way, IoT technology-based applications operate over an architecture consisting of a wide variety of heterogeneous devices, ranging from powerful devices (such as servers, smartphones, laptops, and raspberry pi (Rasberrypi, 2022)) to small devices (such as Arduino UNO (Arduino, 2022)) operated by micro-controllers. The robust devices host standard operating systems, implement common protocols and frameworks, and offer powerful multimodal communication capabilities, while small resource-constrained devices exhibit different characteristics such as lightweight operating systems (e.g., TinyOS (TinyOS, 2022), Contiki (Contiki, 2022)) and operate on weaker communication protocols (e.g. Bluetooth, ZigBee).

In addition, IoT-based information systems use different types of protocols (e.g., Low Rate Wireless Personal Area Networks (LR-WPANs) (IEEE, 2012), Low Power Wide Area Network (LPWAN), Low Power Wireless Personal Area Network (LoWPAN), Internet Control Message Protocol (ICMP), and Constrained Application Protocol (CoAP) (Shelby et al., 2014) for applications operation purpose.

Therefore, the entire deployment of IoT architecture requires to be protected from attacks that may obstruct the services provided by the automated information system. Firstly, it may intimidate security and privacy-related issues of enterprise operational data. Secondly, since IoT-based information systems deal with different interconnected types of networks and many categories of devices, it poses security threats to the whole system. In addition, resource constraints IoT devices demand special security management techniques. Consequently, IoT-based information systems need special security measures.

IoT SECURITY REQUIREMENTS AND CHALLENGES

In general, whenever there is an asset which is valuable to someone (or an organization), there is a need for security measures. Valuable assets could be 'large' sums of money, legal or sensitive documents, a list of secret medical records, police records, students' marks before its announcement, and a new car design before its launch. Before the arrival of the computer and digital age, most security systems depended on safes with lockers and keys disguised and placed under *guards* and *traps*. Such systems' sophistication and ingenuity depend on the value of the object to be guarded and the seriousness of attacks. Within this context, security relates to (i) controlling access to the valuable object, (ii) depriving the *"enemy"* of any valuable information, (iii) protecting sensitive secrets, and (iv) preventing the falsification of records.

Identifying those who should have access and recognizing the enemy was easy. Advances in computer technology change the meaning and scope of security. However, unlike the security situation in the physical world, adding more security tools and hurdles to a computer system does not necessarily enhance its security. In order to understand security in the digital world, one needs to have a look at the way security issues arose in this context.

Since the dawn of electronic computing, there have been rapid advances in computer technology and communication technology. These two technologies have converged and become intertwined in recent decades, affecting every aspect of human activity and business operations. At the same time, the use of computers led to an increased capability in the data processing. Large databases have been created to hold sensitive and not too sensitive information. In the beginning, access to such databases was limited to certain *authorized users*, often known to the system operators. At that stage, passwords were deemed enough to guard against unauthorized users. Nevertheless, as time-sharing and remote access became more widespread, exposure to threats increased dramatically.

Once IoT technology-driven information systems connected using data communication channels and large networks arrived on the scene, remote access became available to many users. This led to an explosion in the amount of information

Figure 4. Categories of threats

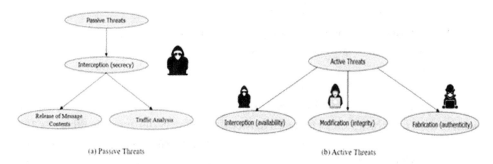

(a) Passive Threats   (b) Active Threats

that could be gathered, stored, and accessed remotely, bringing new exciting and excellent benefits. At the same time, opportunities for misuse became larger.

THREATS TO IoT-BASED INFORMATION SYSTEMS

Attacks on IoT-based information systems (e.g., computer systems/networks) target the computer as an information provider. There are basic requirements for computer systems from security perspectives: (i) secrecy – only authorized users have access, (ii) integrity – only authorized users can make changes, and (iii) availability – the assets are available for authorized users.

However, the unauthorized user (or hacker) can create problems for the data in an information system. The diverse categories of problems are – (i) interruption (i.e., prevents availability), (ii) interception (i.e., breaks the secrecy or confidentiality of the data), (iii) modification (i.e., attacks the integrity), and (iv) fabrication (i.e., attacks the authenticity). In addition, some research literature broadly classified information systems security threats into two classes – passive treats and active threats. These threats are diagrammatically shown in Figure 4.

In a computerized information system, data flow from a source (file/region of memory) to a destination (file/user). Hence, one can identify four general types of attacks, shown in Figure 5, and discussed consequently.

- **Interruption**: This attacks the availability of computer assets that are either destroyed or made unusable. Examples include cutting a computer network or disabling the data management activity.
- **Interception:** This is an attack on confidentiality. Examples include wiretapping and copying files or programs.

- **Modification:** This attacks integrity; examples include chancing records in files or changing the contents of a message.
- **Fabrication:** Fabrication attacks authenticity. Examples include adding records to a data storage or interesting messages in a network (Nawir et al., 2016).

Attacks that aim to obtain information and monitor transmission are classified as *passive attacks*, while attacks that aim to modify data streams or create false streams are known as *active attacks*. There are various sources of attacks on a system, and they can create several types of problems. The following section presents a few examples.

Figure 5. Various forms of threats

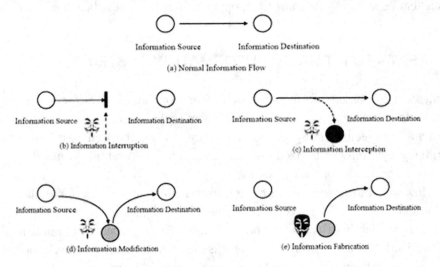

- **Denial of Service Attack** (DOS): An attempt to make an IoT-based information system's resource unavailable to its intended users. DOS attack in the physical layer causes jamming (an unauthorized party occupies and disables computer networks) and node tampering (valuable business information is taken away by physical hardware tampering). These attacks cause spoofing (a useless message sent by a malicious network node that the attacker then replays to produce a massive volume of traffic) (Ali-Shah et al., 2017).

PROTECTION MECHANISMS FOR IoT-Based INFORMATION SYSTEMS

In general, IoT-based information systems aim to provide:

- **Secrecy and Confidentiality**: Information in an IoT-based computer system or on the route should only be accessible for reading by authorized parties. Secrecy is a term which refers to the effect of mechanisms used to limit the number of authorized principals who can access the information. Confidentiality involves an obligation to protect someone else's secrets if an individual knows them; for example, a transaction between two parties cannot be accessed by anyone else. It includes authentication and identity management (Mahalle et al., 2010).
- **Integrity:** Data and information assets in IoT-based information systems can only be modified by authorized parties. This also means that information must remain the same as it was intended. For example, messages received at the end of a transaction have not been tampered with in transit.
- **Availability:** IoT-based enterprise information systems assets or services are available to authorized parties whenever and wherever needed.

Ideally, such a system should also be able to: (i) distinguish between an attacker and a genuine authorized user, (ii) spot unauthorized modification to a document, and (iii) recognize that an attack has taken place.

The use of the IoT-based information system for commercial transactions and the emergence of multimedia objects and mobile computing devices have led to new forms of threats to digital information assets. Besides the above-stated aims of security mechanisms, the general aim of protection for the new economy includes:

- **Authenticity**: Participants in a transaction are assured of each other's identity, even though they may have never met and there is no physical presence. This way, authentication and identity management play a vital role in a thriving IoT environment (Mahalle et al., 2010)
- **Non-repudiation**: A message sender cannot claim later that it did not send it. This aim relates to protection against fraud, such as the case with cards, not a present problem that costs banks vast sums of money.
- **Privacy**: The ability and right to protect some entity's secrets (or sensitive information). It can also extend to families/groups but not to legal entities such as corporations. For example, in an IoT environment, a legal entity needs to be uniquely identified and traceable in communication (Borgohain et al., 2015).

- **Ownership of digital objects**: Ownership of digital objects store and share sensitive information over the IoT network, and they might be vulnerable to unauthorized predators, and valuable data could be stolen (Whitmore et al., 2015).

In order to make IoT services available at low cost, with many devices communicating securely with each other, there are many security challenges to overcome. Some primary challenges are: (i) scalability – managing many IoT nodes requires scalable security solutions, (ii) connectivity – in IoT communications, connecting various devices of different capabilities securely is another challenge, and (iii) end-to-end security – it measures the protection of data leakage between IoT devices and Internet hosts, and other categories of threats. This includes security and data protection, authentication and identity management, and privacy-related issues.

With the introduction of new technologies, some of the weaknesses of IoT-based systems' privacy security can be managed with different solutions. In the following sections, two aspiring technology paradigms are described with some quick solutions for IoT systems: cryptography-based solutions and blockchain-based solutions.

CRYPTOGRAPHY FOR THE INTERNET OF THINGS

Cryptography is a technique for dealing with IoT-based information system data in a transformed form that can only be read and processed by an intended group. It is the science of securing data and information by changing it into an unrecognizable form. This way, it is an appropriate mechanism for protecting business-sensitive information stored on media or through data communication networks. It has long been recognized that disguising the message by scrambling it prior to storage or transmission may provide the required protection. This procedure is known as *encryption*, and the inverse procedure of recovering a message from its encrypted form is called *decryption*. In this way, cryptography is the study of *encryption* and *decryption* systems. It is worth understanding the following technical terms in order to understand the basics of cryptography technique:

- **Cryptography:** The science of keeping data secure. Cryptography is different from steganography used to conceal a message by hiding it so that its mere existence is obscured.
- **Encryption:** An algorithm to scramble messages before storage or transmission to make them unintelligible to eavesdroppers. Other terms used for encryption include enciphering and encoding.

- **Decryption:** An algorithm which aims at transforming an encrypted message back to its form before encryption. Decryption is also referred to as deciphering or decoding. For example, a decryption algorithm has matching encryption.

That includes both encryption and matching decryption processes and its matching decryption process.

- **Cryptoanalysis:** The art and science of breaking cryptosystems. A cryptanalyst attempts to deduce the meaning of encrypted messages without complete knowledge of the decryption process or determine a decryption algorithm that matches an encrypting one.
- **Cryptology:** The study of both cryptography and cryptoanalysis.

The context of IoT-based information security systems has a few components that mimic those used for physical security:

1. **keys** – to safeguard the secrecy of messages and files,
2. **efficient algorithm** – to lock or unlock the document with the appropriate key, and
3. **communication protocols** – to safeguard the passage of messages and files.

Given a language L with a finite alphabet αL. For example, one can only consider the English language. For convenience, one can consider limiting its alphabet to {A, B, ..., Z}. For computers, L has preferably represented in ASCII, Binary, or object codes. Most texts use the simplified integer representation of the English alphabet, i.e., $A = 0, B = 1, C = 2, ..., Z = 25$. It is worth explaining three main components:

1. **Message** – a message in L is just a string of characters in L. The set of all messages in L is denoted by L^*.
2. **Plaintext** – it is an unencrypted message in L, and often this means that it is meaningful.
3. **Ciphertext** – it is an encrypted message in L, i.e., a message whose content is concealed with the aid of an encryption algorithm.

This section will describe a simple cryptosystem known as Julius Caesar's Security system with the above technical definitions. Julius Caesar invented this system to communicate with his commanders. The individual character of a message is replaced by the character 3 characters ahead in the alphabet. The following diagram in Figure 6 illustrates the English equivalent.

For example, "HELLO WORLD" encrypts to "KHOOR ZRUOG," and "FDHVDU" decrypts to "CAESAR". Using the integer representation of the English characters, this encryption can be achieved numerically by using the following equation:

$$y = x + 3 \text{ Mod } 26$$

Here, x is the numerical of the character to be encrypted, and y is the numerical value of the corresponding ciphertext. Moreover, decryption can be achieved by the modular equation:

$$y = x + 23 \text{ Mod } 26$$

Figure 6. Diagrammatic representation of a cryptosystem

Caesar's system can be generalized in many ways. The most straightforward such generalization is trying to use a shift of any size instead of the fixed shift of 3.

A shift substitution cryptosystem has a key b, which is an integer in the range 1 ... 25; encryption is achieved by the equation:

$$y = x + b \text{ Mod } 26$$

and decryption is achieved by the equation

$$y = x + b \text{ Mod } 26$$

where B is an integer in the range 1 ... 25 such that b + B = 26. The integer B is called the additive inverse of b. For example, if b = 5, then "HELLO WORLD" encrypts to "MJQQT BTWQI" and "HFJXFU" decrypts to "CASAR".

A more general form is provided by *affine substitution*. An affine substitution cryptosystem has a key which is a pair (a, b) of integers in the range 1...25 such that a is relatively prime to 26, (i.e., has no common factor with 26). The following equation achieves encryption:

$$y = ax + b \text{ Mod } 26$$

and decryption is achieved by the equation

$$y = Ax + B \text{ Mod } 26$$

where A is an integer in the range 1..25, such that aA = 1 Mod 26, and B is the additive inverse of Ab (i.e., Ab + B = 0 mode 26). The integer A is called the multiplicative inverse of a and is customary to denote it by a^{-1}. Once A is determined, B is evaluated by the formula:

$$B = 26 - (A*b \text{ Mod } 26)$$

For example, if a = 5 and b = 3 then "HELLO WORLD" encrypts to "MXGGV JVKGS" and "NDXPDK" decrypts to "CAESAR". For the corresponding decryption formula: A = 21, and B = 26 – (A*b Mod 26) = 26 – (63 Mod 26) = 26 -11 = 15, i.e., the decryption key pair is (21, 15).

The modular arithmetic calculations of affine ciphers can be simplified, so the same lookup table performs that encryption and decryption. This follows from the observation that, for an affine cipher with a key pair (a, b), character A is encrypted to the one with numerical value b and the encryption of any other character can be obtain from that of its predecessor, in the alphabet, by a shift of value a. Consequently, a lookup table can be constructed by iteration.

For example, if a = 5 and b = 3, then the lookup table is as follows:

Affine substitution ciphers, including Caesar, have the following properties: (i) very simple and can be performed by a lookup table, (ii) belong to a class of systems called *mono-alphabetic*, i.e., each character of the alphabet is enciphered to the same character regardless of its position within the plaintext, and hence (iii) the frequency distribution of the ciphertext characters reflects that of the plaintext characters, and (iv) consequently, this cipher amenable to statical analysis attacks.

Table 1.

Character	A	B	C	D	E	F	G	H		O	P	Q	R	S	T	U	V	W	X	Y	Z
Number value	0	1	2	3	4	5	6	7		14	15	16	17	18	19	20	21	22	23	24	25
Encryption value	3	8	13	18	23	2	7	12		21	0	5	10	15	20	25	4	9	14	19	24

Classification of Modern Ciphers

There are two ciphers in modern cryptography: (i) private-key cryptography and (ii) public-key cryptography.

Private Key Cryptography

A private key cipher is one whose encryption and decryption keys are kept secret. All classical ciphers belong to this class. Modern private key ciphers, also known as conventional, use the same key for encryption and decryption (i.e., symmetric). Conventional ciphers are further divided according to how they work: (i) stream ciphers – text is processed a character at a time, and (ii) block ciphers – text processes in fixed-size blocks. Figure 7 describes the steps necessary to create private key cryptography.

Figure 7. Different steps in a private-key cryptography

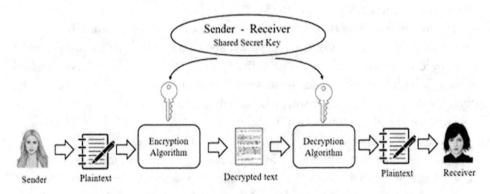

The following is a list of main characteristics of modern private key ciphers: (i) messages are transformed into bit-strings, and then encryption/decryption work by bit map operations and manipulations involving the secret key, (ii) a secure n users communication private key system needs n(n – 1)/2 keys which are random

Figure 8. Strategic steps in a public-key cryptography

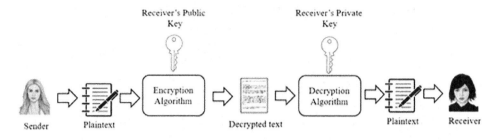

bit strings, (iii) exhaustive search is the only theoretical way to obtain the key; and larger keyspace, in general, mean more secure ciphers, and (iv) the secret key requires to be exchanged over a "secure" channel.

Private key ciphers are expected to produce the following effects on the plaintext: (i) *confusion* – information (i.e., plaintext) is changed so that the output bits have no apparent relation to the input bits, and (ii) diffusion – the effect of one plaintext bit is spread to as many other bits in the ciphertext. These concepts are based on Shannon's Theory of Information security (Shannon, 1949).

Public Key Cryptography

A public key cipher is one whose encryption key is made public, but the decryption key is kept secret (i.e., *asymmetric*). Public key systems (initially proposed in the early 1970s) and the following is a list of main characteristics of modern public-key ciphers: (i) public-key algorithms are based on specific mathematical structures (e.g., Abelian groups), (ii) messages are transformed into integers, and then encryption/ decryption work by numerical operations in the underlying mathematical structures, and (iii) a secure n-user network private key system needs only *n* pairs; and there is no need to exchange keys. Figure 8 describes a diagrammatic representation of the public-key cryptography technique.

For secure public-key ciphers, deducing the private key from the public key is assumed to be equivalent to solving a computationally *'hard'* problem.

CRYPTOGRAPHY RESEARCH WORKS

The improvement of the new generation of RFID technology presents a new era of ubiquitous computing and machine perception. In this way, ubiquitous and pervasive computing demands a particular type of cryptography, known as lightweight cryptography, for deployed infrastructure security purposes. Keeloq (Bogdanov,

2007) is the first lightweight block cipher. Lightweight block ciphers need the help of simple operations such as bitwise XOR, bitwise AND, and Substitution box (S-box), leading to an increased required number of rounds. For security purposes, the lightweight ciphers require appropriate cryptographic algorithms.

Hardware Implementation

Lightweight cryptography hardware implementation is based on predefined characteristics, such as code size, random-access memory (RAM) utilization, and energy consumption, which are the essential attributes in the design process. In computational hardware industrial industries, lightweight cryptography evaluation often plays an essential role, which considers the implicational hardware circuit (e.g., clock), memory (e.g., RAM), storing the internal states and key states is vital to be considered for design purposes. Again, other deciding properties are block and key size. In recent decades, a unique hardware implementation metric has been selected for the energy efficiency of cryptography operation.

Software Implementation

Lightweight cryptography software deployment is often considered a preferable metric that consists of implementation size, RAM use in the operation, and throughput (bytes per cycle) for application development. The simple design criteria are that the small, the better. It also considers the performance based on implementation size, RAM utilization, and time is taken to perform a particular operation (Dinu et al., 2015). Table 2 shows a comparison of some lightweight block chippers.

Different reviewers presented various aspects of cryptography-based techniques for resource constraints in IoT-based information systems. For example, a research group examined resource constraint IoT device-related cryptographic methods for types of information systems-related issues (Bossuet et al., 2013), and some other researchers presented their findings for embedded platforms (Law et al., 2006) (Eisenbarth et al., 2012). In addition, some research groups mainly dedicated their research efforts to cryptography methods for hardware implementation (Kerckhof et al., 2012) (Batina et al., 2013) and other groups looked at software design-related issues (Cazorla et al., 2013) (Weis & Lucks, 2000). At the same time, a few research groups try to assess the performance of cryptographic methods for both hardware and software related design issues (Eisenbarth et al., 2007). Figure 7 represents a comparison between different categories of lightweight block ciphers.

Cipher hardware implementations have recently received intensive research contributions. Many resource constraint ciphers were introduced in recent decades; some are listed in Table 2. Several studies were conducted to compare the ciphers;

Table 2. Design description of some standard lightweight ciphers

Ciphers	Function	Architecture	Structure	Key size	Block size	Rounds	Cycles / Block	Throughput	Area (GEs)
PRINT-48	Enc	Serialized	SPN	80	48	48	768	6.25	402
PRINT-48	Enc	Round-based	SPN	80	48	48	48	100	503
PRINT-96	Enc	Serialized	SPN	160	96	96	3072	3.11	726
PRINT-96	Enc	Round-based	SPN	160	96	96	96	100	967
LED-64	Enc	Serialized	SPN	64	64	32	1248	5.1	688
LED-80	Enc	Serialized	SPN	80	64	48	1872	3.4	690
LED-96	Enc	Serialized	SPN	96	64	48	1872	3.4	695
LED-128	Enc	Serialized	SPN	128	64	48	1872	3.4	700
KTANTAN-32	Enc	Serialized	LFSR	80	32	254	255	12.5	462
KTANTAN-48	Enc	Serialized	LFSR	80	48	254	255	18.8	588
KTANTAN-64	Enc	Serialized	LFSR	80	64	254	255	25.1	688
PRESENT-80	Enc	Serialized	SPN	80	64	32	516	12.4	1030
PRESENT-80	Enc	Round-based	SPN	80	64	32	32	200	1570
PRESENT-128	Enc	Serialized	SPN	128	64	32	528	12.12	1339
PRESENT-128	Enc	Round-based	SPN	128	64	32	32	200	1886
EPCBC-48	Enc	Serialized	SPN	96	48	32	396	12.12	1008
EPCBC-96	Enc	Serialized	SPN	96	96	32	792	12.12	1333
DES	Enc	Serialized	Feistel	56	64	16	144	5.55	2309
DESL	Enc	Serialized	Feistel	56	64	16	144	5.55	1848
DESX	Enc	Serialized	Feistel	184	64	16	144	5.55	2629
DESXL	Enc	Serialized	Feistel	184	64	16	144	5.55	2168
TWINE-80	Enc	Round-based	Feistel	80	64	36	36	178	1503
TWINE-80	Enc	Serialized	Feistel	80	64	36	540	11.8	1116
TWINE-80	Enc+Dec	Round-based	Feistel	80	64	36	36	178	1799
TWINE-128	Enc	Round-based	Feistel	128	64	36	36	178	1866
TWINE-128	Enc+Dec	Round-based	Feistel	128	64	36	36	178	2285
Piccolo	Enc+Dec	Round-based	SPN	128	64	32	32	194	2537
KLEIN-64	Enc+Dec	Round-based	SPN	64	64	12	13	492.3	2475
KLEIN-80	Enc+Dec	Round-based	SPN	80	64	16	17	376.5	2629
KLEIN-96	Enc+Dec	Round-based	SPN	96	64	20	21	304.8	2769
KLEIN-64	Enc+Dec	Serialized	SPN	64	64	12	207	30.9	1220
KLEIN-80	Enc+Dec	Serialized	SPN	80	64	16	271	23.6	1478
KLEIN-96	Enc+Dec	Serialized	SPN	96	64	20	335	19.1	1528
KATAN-32	Enc	Serialized	LFSR	80	32	254	255	12.5	802
KATAN-48	Enc	Serialized	LFSR	80	48	254	255	18.8	927
KATAN-64	Enc	Serialized	LFSR	80	64	254	255	25.1	1054
LED-64	Enc	Serialized	SPN	64	64	32	1248	5.1	966
LED-80	Enc	Serialized	SPN	80	64	48	1872	3.4	1040
LED-96	Enc	Serialized	SPN	96	64	48	1872	3.4	1116
LED-128	Enc	Serialized	SPN	128	64	48	1872	3.4	1265
LBLOCK	Enc	Round-based	Feistel	80	64	32	32	200	1320
LBLOCK	Enc	Serialized	Feistel	80	64	32	576	11.1	866
RECTANGLE-80	Enc	Round-based	SPN	80	64	25	26	246	1467
RECTANGLE-128	Enc	Round-based	SPN	128	64	25	26	246	1787
RECTANGLE-80	Enc	Serialized	SPN	80	64	25	461	13.9	1066

however, older studies did not include comparisons for the most recent ones. For example, figure 9 presents a comparison of a few old and some newly introduced ciphers.

Some of the main research outcomes are as follows: (i) the design of lightweight cipher round contains small combinatorial logic; as a result, with a fixed frequency, increasing the number of unrolled rounds does not directly impact timing delays, and increases area linearly, and (ii) the energy per bit is dependent on several factors; consequently complex key generation reduce energy efficiency, and increasing key-size, and to a lesser extent, reducing block size degrades energy efficiency.

Figure 9. A comparison between different ciphers
LEGEND: A: EPCBC-48, B: KATAN-48, C: KATAN-64, D: LED-80, E: PRESENT-80, F: RECTANGGLE-80
G: LED-80, H: LBLOCK, I: EPCBC-96, J: TWINE-80, K: PRESENT-128, L: LED-128

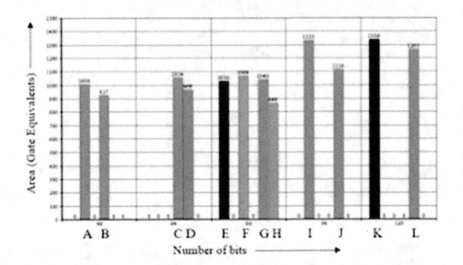

SECURITY SOLUTIONS BY BLOCKCHAIN TECHNOLOGY

Blockchain technology is the distributed ledger technology underlying bitcoin and other cryptocurrencies (Nakamoto, 2008). However, its usage is dramatically expanding beyond the financial industry into other industrial applications (Pal & Yasar, 2020b). Potentially as revolutionary as the Internet and artificial intelligence, blockchain promises disruptive innovations in technology, business models, and industrial automation (Pal, 2022a) (Pal, 2022b). It provides a trustless computing infrastructure for large-scale applications that has the potential to solve fundamental trust issues and allow economic transactions to take place without intermediaries.

In simple, blockchain technology consists of a distributed database system. By creating a trustless systems integration environment, blockchain permits stable, robust, algorithm-ensured data storage and transaction authorization and autonomous processing. Academics and practitioners are hopeful and researching blockchain technology for its role in managing, particularly preserving the security of IoT-based information systems. This way, it uses smart contract techniques to manage better and control IoT devices. Besides, blockchain technology uses elliptic curve cryptography (ECC) and SHS-256 hashing techniques to enact integrity and data authentication for industrial information systems. Table 3 presents some blockchain-based IoT industrial solutions.

Table 3. IoT-based information system and blockchain technology solutions

Security issue	Proposed solutions	References
Data authentication and integrity	*Description of cryptographic-based ECC and SHA-256 hashing to accommodate in information systems security.*	*(Antonopoulos, 2014)*
Ownership and identity relationships of IoT devices	*Blockchain-based solution techniques for facilitating trustworthy and authorized identity registration.*	*(Friese et al., 2014) (Mahalle et al., 2010) (Otte et al., 2017)*
Management of access and control to data	*Immutable log of events and management of access control to data*	*(Conoscenti et al., 2016) (Zyskind et al., 2015)*
Trading of IoT data	*Issues of trading of collected IoT-based information systems data*	*(Zhang & Wen, 2015) (Worner & von Bomhard, 2014)*
IoT device identity management	*Technical issues of symmetric and asymmetric key management for IoT devices*	*(Axon, 2015) (Formkecht et al., 2014)*
Industrial IoT application	*Cloud service and blockchain-based industrial IoT (IIoT) application.*	*(Bahga & Madisetti, 2016)*
Smart contracts	*Industrial applications of blockchain smart contracts to IoT-based applications.*	*(Christidis & Devetsikiotis, 2016) (Pureswaran & Brody, 2014)*

FUTURE WORK

There are essential factors that must be adhered to when comparing the different ciphers to evaluate the performance when implementing the primitives to make them a fair comparison (e.g., security level, design technology). Smart and light technologies are modern trends for industrial applications. Consequently, researchers have been developing and proposing a range of cryptographic algorithms to suit lightweight devices (e.g., RFID tags and sensors in embedded systems). In addition, world standardization authorities (e.g., NIST) provide more LWC research directions. These will help to standardization of lightweight cryptographic algorithms.

With the evaluation of 6G-based devices and data communication technologies, the industry updates IoT device design practices to upgrade to compatible high-speed computer network connectivity. In addition, a detailed review of lightweight cryptographic deployment-related issues, mitigating blockchain technology's security risks and industry-specific solutions will be on the agenda of future research.

CONCLUSION

In Today's world, IoT technology is used immensely in every industrial field like smart home, intelligent transport, and healthcare. The main objective of this chapter is to present an overview concept of IoT technology and its contribution to industries enterprise information system. In the field of healthcare, IoT is brought into use for applications like monitoring of the patient's health condition regularly, drug traceability, and so on. However, there exist various security related issues in IoT, which can be solved by integrating IoT with blockchain technology, and the use cryptography techniques to secure the enterprise data. The blockchain is a decentralized technology that can be used to enhance the security of the information system. Blockchain technology along with healthcare ensures that patients' sensitive health-related records remain safe any type of tampering and leakage.

At the same time, the advancement of industrial IoT technologies provides scalable and less expensive interconnected sensor networks providing the infrastructures to collect, store, and process the massive amount of data for business analytics deployment purposes. Artificial Intelligence's recent popularity is fueled by machine learning breakthroughs, particularly the new generation of machine learning paradigms. In this way, industrial big data applications often deal with unstructured data, which increases the difficulty of searching for meaningful insight. Besides, IoT devices play an essential role in the modern era when conventional devices become more autonomous and intelligent. At the same time, high-speed data transfer is a significant issue where the 5 G-enabled industrial environment plays a prominent role and may cause several security issues for the system applications.

Despite providing helpful IoT systems security solutions, blockchain technology has weaknesses in handling security. For example, the consensus method of blockchain technology depends heavily on miners' hashing power, which could be attacked, allowing predators to manage the blockchain's operational activities. Similarly, private keys with limited randomness can be utilized to compromise the blockchain accounts. Appropriate methods need to be used to ensure the privacy and security of the IoT-based system operations.

REFERENCES

Aazam, M., & Huh, E. N. (2014). Fog computing and smart gateway-based communication for cloud of things. *Proceedings of the 2nd IEEE International Conference on Future Internet of Things and Cloud (FiCloud' 14)*, 464–470. 10.1109/FiCloud.2014.83

Abduvaliyev, A., Pathan, A. S. K., Zhou, J., Roman, R., & Wong, W. C. (2013). On the vital areas of intrusion detection systems in wireless sensor networks. *IEEE Communications Surveys and Tutorials*, *15*(3), 1223–1237. doi:10.1109/SURV.2012.121912.00006

Alaba, F. A., Othman, M., Hashem, I. A. T., & Alotaibi, F. (2017). Internet of things security: A survey. *Journal of Network and Computer Applications*, *88*, 10–28. doi:10.1016/j.jnca.2017.04.002

Ali-Shah, P. A., Habib, M., Sajjad, T., Umar, M., & Babar, M. (2017). Applications and Challenges Faced by Internet of Things - A Survey, in Lecture Notes of the Institute for Computer Sciences. *Social Informatics and Telecommunications Engineering Future Intelligent Vehicular Technologies, Springer*, *2017*, 82–188.

Antonopoulos, A. M. (2014). *Mastering Bitcoin: Unlocking Digital Crypto-Currency* (1st ed.). O'Reilly Media, Inc.

Arduino. (2022). https://www.arduino.cc/en/main/arduinoBoardUno

Axon, L. (2015). *Privacy-awareness in Blockchain-based PKI*. Technical Report.

Bahga, A. & Madisetti, V.K. (2016). *Blockchain platform for industrial Internet of Things*. Technical Report.

Bonomi, F., Milito, R., Natarajan, P., & Zhu, J. (2014). *Fog computing: a platform for internet of things and analytics. In Big Data and Internet of Things: A Road Map for Smart Environments*. Springer.

Bonomi, F., Milito, R., Zhu, J., & Addepalli, S. (2012). Fog computing and its role in the internet of things. *Proceedings of the 1st ACM MCC Workshop on Mobile Cloud Computing*, 13–16. 10.1145/2342509.2342513

Borgia, E. (2014). The Internet of Things vision: Key features, applications, and open issues. *Computer Communications*, *54*, 1–31. doi:10.1016/j.comcom.2014.09.008

Borgohain, T., Kumar, U., & Sanyal, S. (2015). Survey of Security and Privacy Issues of Internet of Things. *International Journal of Advanced Network Applications*, *6*(4), 2372-2378.

Butun, I., Morgera, S. D., & Sankar, R. (2014). A survey of intrusion detection systems in wireless sensor networks. *IEEE Communications Surveys and Tutorials*, *16*(1), 266–282. doi:10.1109/SURV.2013.050113.00191

Christidis, K., & Devetsikiotis, M. (2016). Blockchains and Smart contracts for the Internet of Things. *IEEE Access: Practical Innovations, Open Solutions*, *4*, 2292–2303. doi:10.1109/ACCESS.2016.2566339

Cirani, S., Ferrari, G., & Veltri, L. (2013). Enforcing security mechanisms in the IP-based internet of things: An algorithmic overview. *Algorithms*, *6*(2), 197–226. doi:10.3390/a6020197

Conoscenti, M., Vetro, A., & Martin, J. C. D. (2016). Blockchain for the Internet of Things: A systematic literature Review. *3rd International Symposium on Internet of Things: Systems, Management, and Security, IOTSMS-2016.*

Contiki-OS. (2022). http://www.contiki-os.org/

Dinu, D., Biryukov, A., & Großschädl, J. (2015). FELICS – fair evaluation of lightweight cryptographic systems. In *NIST Workshop on Lightweight Cryptography*. National Institute of Standards and Technology (NIST).

Friese, I., Heuer, J., & Kong, N. (2014). Challenges from the Identities of Things: Introduction of the definitive of Things discussion group with Kantara initiative. *IEEE World Forum on Internet of Things (WE-IoT)*, 1-4.

Fromknecht, C., Velicanu, D., & Yakoubov, S. (2014). *CertCoin: A namecoin based decentralized authentication system*. Academic Press.

Granjal, J., Monteiro, E., & Silva, J. S. (2015). Security for the internet of things: A Survey of existing protocols and open research issues. *IEE Communication Surveys and Tutorial, 17*(3), 1294–1312. doi:10.1109/COMST.2015.2388550

Granjal, J., Silva, R., Monteriro, E., Silva, J. S., & Boavida, F. (2008). Why is IPSec a viable option for wireless sensor networks. *5th IEEE International Conference on Mobile Ad Hoc and Sensor Systems*, 802-807. 10.1109/MAHSS.2008.4660130

Gubbi, J., Buyya, R., Marusic, S., & Palaniswami, M. (2013). Internet of Things (IoT): A vision, architectural elements, and future directions. *Future Generation Computer Systems, 29*(7), 1645–1660. doi:10.1016/j.future.2013.01.010

IEEE. (2012). https://ieeexplore.ieee.org/document/6185525

Internet of Things–Architecture. (2013). *IoT-A Deliverable D1.5 – Final architectural reference model for the IoT v3.0*. Academic Press.

Kerckhof, S., Durvaux, F., Hocquet, C., Bol, D., & Standaert, F. X. (2012). *Towards green cryptography: a comparison of lightweight ciphers from the energy viewpoint*. In *Cryptographic hardware and embedded systems–CHES 2012*. Springer.

Law, Y., Doumen, J., & Hartel, P. (2006). Survey and benchmark of block ciphers for wireless sensor networks. *ACM Transactions on Sensor Networks, 2*(1), 65–93. doi:10.1145/1138127.1138130

Mahalle, P., Babar, S., Prasad, N. R., & Prasad, R. (2010). Identity Management Framework towards Internet of Things (IoT): Roadmap and Key Challenges. In *Recent trends in network security and applications, communications in computer and information science*. Springer.

Mitchell, R., & Chen, I. R. (2014). Review: A survey of intrusion detection in wireless network applications. *Computer Communications, 42*, 1–23. doi:10.1016/j.comcom.2014.01.012

Nakamoto, S. (2008). *Bitcoin: A peer-to-peer electronic cash system*. Academic Press.

Nawir, M., Amir, A., Yaakob, N., & Lynn, O. B. (2016). Internet of Things (IoT): Taxonomy of security attacks. *3rd International Conference on Electronic Design*, 321-326. 10.1109/ICED.2016.7804660

Nodered. (2022). https://nodered.org/

Oleshchuk, V. (2009). Internet of things and privacy preserving technologies. *2009 1st International Conference on Wireless Communication, Vehicular Technology, Information Theory and Aerospace Electronic Systems Technology*, 336–340.

Otte, P., de Vos, M., & Pouwelse, J. (2017). TrustChain: A Sybil-resistant scalable blockchain. *Future Generation Computer Systems*.

Pal, K. (2019). Algorithmic Solutions for RFID Tag Anti-Collision Problem in Supply Chain Management. *Procedia Computer Science*, 929-934.

Pal, K. (2021a). Privacy, Security and Policies: A Review of Problems and Solutions with Blockchain-Based Internet of Things Applications in Industrial Industry. *Procedia Computer Science*.

Pal, K. (2021b). A Novel Frame-Slotted ALOHA Algorithm for Radio Frequency Identification System in Supply Chain Management. *Procedia Computer Science*, 871-876. 10.1016/j.procs.2021.03.110

Pal, K. (2022a). Application of Game Theory in Blockchain-Based Healthcare Information System. In Prospects of Blockchain Technology for Accelerating Scientific Advancement in Healthcare. The IGI Global Publishing.

Pal, K. (2022b). Semantic Interoperability in Internet of Things: Architecture, Protocols, and Research Challenges. In Management Strategies for Sustainability, New Knowledge Innovation, and Personalized Products and Services. The IGI Global Publishing.

Pal, K. (2022b). A Decentralized Privacy Preserving Healthcare Blockchain for IoT, Challenges and Solutions. In Prospects of Blockchain Technology for Accelerating Scientific Advancement in Healthcare. The IGI Global Publishing.

Pal, K., & Yasar, A. (2020b). Semantic Approach to Data Integration for an Internet of Things Supporting Apparel Supply Chain Management. *Procedia Computer Science,* 197 - 204.

Pal, K., & Yasar, K. (2020a). Internet of Things and Blockchain Technology in Apparel Manufacturing Supply Chain Data Management. *Procedia Computer Science,* 450 - 457.

Pureswaran, V., & Brody, P. (2014). *Device Democracy – Saving the future of the Internet of Things.* IBM.

Raspberrypi. (2022). https://www.raspberrypi.org

Roman, R., Alcaraz, C., Lopez, J., & Sklavos, N. (2011). Key management systems for sensor networks in the context of the internet of things. *Computers & Electrical Engineering, 37*(2), 147–159. doi:10.1016/j.compeleceng.2011.01.009

Roman, R., Lopez, J., & Mambo, M. (2016). Mobile edge computing. In *A survey and analysis of security threats and challenges.* Future Generation Computer Systems.

Said, O. & Masud, M. (2013). Towards Internet of Things: Survey and Future Vision. *International Journal of Computer Networks, 5*(1), 1–17.

Shannon, C. E. (1949). Communication Theory of Secrecy Systems. *The Bell System Technical Journal, 28*(4), 656–715. doi:10.1002/j.1538-7305.1949.tb00928.x

Shelby, K., Hartke, K., & Bormann, C. (2014). *The constrained application protocol (CoAP).* https://tools.ietf.org/html/rfc7252

Sicari, S., Rizzardi, A., Grieco, L., & Coen-Porisini, A. (2015). Security, privacy and trust in internet of things: The road ahead. *Computer Networks, 76*(Suppl. C), 46–164. doi:10.1016/j.comnet.2014.11.008

Stojmenovic, I., & Wen, S. (2014). The fog computing paradigm: scenarios and security issues. *Proceedings of the Federated Conference on Computer Science and Information Systems (FedCSIS' 14),* 1–8. 10.15439/2014F503

TinyO. S. (2022). https://en.wikipedia.org/wiki/TinyOS

Wang, Y., Uehara, T., & Sasaki, R. (2015). Fog computing: Issues and challenges in security and forensics. *IEEE 39th Annual Computer Software and Applications Conference, 3,* 53–59.

Weis, R., & Lucks, S. (2000). The performance of modern block ciphers in java. In J.-J. Quisquater & B. Schneier (Eds.), *Smart card research and applications*. Springer.

Whitmore, A., Agarwal, A., & Xu, L. D. (2015). The Internet of Things—A survey of topics and trends. *Information Systems Frontiers, 12*(2), 261-274.

Worner, D., & von Bomhard, T. (2014). When your sensor earns money: Exchanging data for cash with bitcoin. *Proceedings of the 2014 ACM International Joint Conference on Pervasive and Ubiquitous Computing: Adjunct Publication*, 295-298.

Yi, S., Qin, Z., & Li, Q. (2015). Security and privacy issues of fog computing: A survey. *10th International Conference on Wireless Algorithms, Systems, and Applications*, 1–10.

Zhang, Y., & Wen, J. (2015). An IoT electric business model based on the protocol of bitcoin. *18th International Conference on Intelligence in Next Generation Network*, 184-191.

Zhang, Z. K., Cho, M. C. Y., Wang, C. W., Hsu, C. W., Chen, C. K., & Shieh, S. (2014). IoT security: Ongoing challenges and research opportunities. *7th IEEE International Conference on Service-Oriented Computer Applications*, 230–234.

Zhou, J., Cao, Z., Dong, X., & Vasilakos, A. V. (2017). Security and privacy for cloud-based IoT: Challenges. *IEEE Communications Management, 55*(1), 26–33.

Zyskind, G., Nathan, O., & Pentland, A. (2015). *Enigma: Decentralized computation platform with guaranteed privacy*. Academic Press.

KEY TERMS AND DEFINITIONS

Cryptoanalysis: The art and science of breaking cryptosystems. A cryptanalyst attempts to deduce the meaning of encrypted messages without the complete knowledge of the decryption process, or to determine a decryption algorithm that matches an encrypting one.

Cryptography: The art and science of keeping messages secure. Cryptography is different from steganography which is used to conceal a message by hiding it in such a way that the mere existence is obscured.

Cryptology: The study of both cryptography and cryptoanalysis.

Cryptosystem: A system which includes both encryption process and its matching decryption process.

Decryption: An algorithm which aims at transforming an encrypted message back to its form before encryption. Decryption is also referred to as deciphering or decoding. A decryption algorithm has a matching encryption algorithm.

Encryption: An algorithm with the aim of scrambling messages prior to storage or transmission to make them unintelligible to eavesdroppers. Other terms used for encryption include enciphering and encoding.

Chapter 9

Security Optimization of Resource–Constrained Internet of Healthcare Things (IoHT) Devices Using Lightweight Cryptography

Varsha Jayaprakash
Vellore Institute of Technology, Chennai, India

Amit Kumar Tyagi
ⒾⒹ https://orcid.org/0000-0003-2657-8700
Vellore Institute of Technology, Chennai, India

ABSTRACT

The term "internet of things" is becoming increasingly popular and promising, ushering in a new era of smarter connectivity across billions of gadgets. In the foreseeable future, IoT potential is boundless. The healthcare industry, often known as IoHT, is the most demanding application of IoT. Sensors, RFID, and smart tags are used to start any IoT system, but these applications lack the necessary resources such as power, memory, and speed. The key requirement is secure information transformation because it contains sensitive patient information that might be extremely dangerous if it falls into the hands of an unauthorized person. Encryption approaches that have been used in the past are ineffective. Lightweight cryptography is the most viable solution for protection of data at the physical layer.

DOI: 10.4018/978-1-6684-3921-0.ch009

INTRODUCTION

The Internet of Things has become the most widely used term in the world today. It is a technical concept that entails practical devices such as sensors and actuators that are used to collect real-time data, convey that data over the internet, and store that data on cloud-based platforms with or without human participation (Gubbi et al., 2013; Singh, Sharma, & Moon, 2017; Thakor et al., 2021). In 1999, Kevin Ashton coined the term "Internet of Things" to promote the usage of radio frequency-based identification (RFID), which involves a variety of embedded devices. With the advent of home automation, industrial energy meters, wearable and self-health care devices in 2011, the tremendous expansion of IoT-based devices began (Tawalbeh et al., 2020). Health care is an important sector that is one of the major contributors to the total number of IoT enabled devices in the world. The invention of IoHT enables patients to self-assess their body conditions and also simultaneously upload these data to the hospital's server so that doctors can keep track of patients' health condition and call for checkups and visits only when required which ultimately helps in saving money as well as time (Engineering, 2017; Fuzon, 2019). However, the massive outbreak of this technology has led to many issues and challenges regarding the security of patient's data.

Data protection is required at three layers in any IoHT device: physical/design, communication, and computation. (McKay et al., 2017) They are further divided into resource-rich (phones, tablets, laptops) and resource-constrained (sensors, RFID) devices. Devices with limited resources are frequently utilized to handle real-time applications that demand precise data processing. Furthermore, they are constrained in terms of power consumption, memory, and processing rates (Biryukov & Perrin, 2017; Toshihiko, 2017). The focus of this research is on the implementation of algorithms for device security in the latter group.

In most of the countries, the authentic information provided by the healthcare data should be confined through "Health Information and Portability Accountability Association (HIPAA)" (Ullah et al., 2018). Efficient and safe implementation of these healthcare systems can be achieved by using optimized and robust security systems (Butpheng et al., 2020). Cryptography is the widely applied technique to secure the data and prevent the leakage of information. An IoHT device begins at the implementation of physical layer using sensors, RFID tags, actuators etc. to acquire the information regarding patient's health. Typical encryption algorithms like AES, DES, RSA cannot be applied to these embedded devices as they are more suitable for devices with high computation powers. Lightweight cryptographic (LWC) techniques are utilized in such fields.

As the term suggests they are capable of operating at lower power, smaller memory and better computation speeds (McKay et al., 2017). The most commonly

Figure 1. Categorization of IoT devices

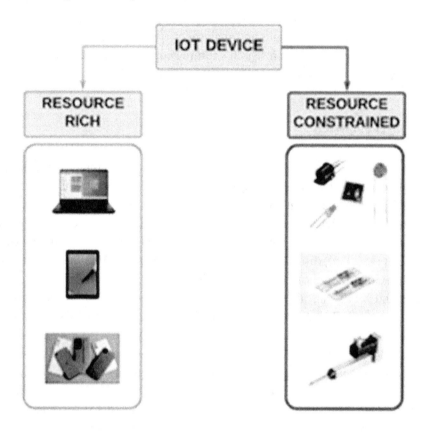

used LWC methods in the field of healthcare are PRESENT, CLEFIA, PICOLO, KATAN, SPECK and SIMON. These ciphers are most widely used in IoHT devices as they are more advantageous in both software as well as hardware application. The function of any cryptography is to convert any plain text to corresponding cipher text using secret keys, logical shifting operations and permutation levels and then convert the cipher text back into plain text. Block ciphers are given more importance in cryptographic fields as they are more efficient and versatile in choosing the information including the key size (Chaudhury et al., 2017; El-hajj et al., 2017).

Organization of the Work

The chapter is organized as follows: Section II give a literature survey regarding related researches followed by motivation in Section III. Section IV discusses the difficulties of developing a secure IoT system, while Section V offers a solution based on the use of various lightweight block cypher algorithms. The working principle and algorithm of the used block cypher cryptography techniques are explained in Section VI. The simulation results and analyses are presented in Section VII. Section VIII gives an overview of the research's future scope, followed by a conclusion in section IX.

RELATED WORKS

Bassam Aboushosha (Aboushosha et al., 2020), proposed a symmetric block cipher technique called SLIM based on 32-bit block size Feistal structure which uses 4 S-boxes to perform nonlinear operation on 16-bit word and is highly efficient against linear and differential attacks. The technique proved to be effective in wireless sensor networks where the transmission width is only few bytes. Further, XinXin Fan (Fan et al., 2013), discusses a lightweight stream cipher WG-8 originated from the Welch-Gong family of cryptography. Some of the block ciphers such as TEA Wheeler and XTEA has also been proposed. Implementation of this cipher in low power microcontrollers proved that they are highly efficient and consume very less power. Then, an ultra-lightweight block cipher called QTL has been suggested by Li et al. (Li et al., 2016) which is slight variation of the FN and is capable of operating at faster speeds compared to other standard internal encryption structures. QTL follows the same encryption and decryption process and is proved to occupy a smaller area and highly cost effective. Further, Biswas (Biswas et al., 2015) surveyed a verity of security mechanism such as KATAN, TWINE, AES and LED which are some standard mechanisms adopted for data confidentiality. He proposed a technique using chaotic maps and genetic operations which uses points on elliptical curves to find the communicating nodes. Moreover this, SecureData, a method developed for IoT based human services that collects data by preserving the privacy of the users was analyzed by Hai Tao (Tao et al., 2018).

The developed method was tested on FPGA equipment using the KATAN technique of encryption. At the cloud level, a circulated database method was adopted in order to preserve patient's privacy. The obtained results proved to be authoritative and authentic. A lightweight blockchain architecture for healthcare database management was proposed by Leila Ismail and Huned Materwala (Ismail et al., 2019). The network participants are divided into demographic clusters by maintaining one copy of ledger. Forking is avoided by using a using a Head Blockchain Manager to handle transactions. The proposed method outperforms traditional Bitcoin network in terms of network traffic generated and computation speed. Further, a novel ultra-lightweight cryptographic technique named Hummingbird was presented by Daniel Engels, Xinxin Fan (Harikrishnan & Babu, 2015) provides security with minimal block size and is efficient against linear and differential attacks. The analysis was performed on 8-bit Atmel and 16-bit Texas instruments microcontrollers. The simulation showed to achieve 4.7 times faster throughput compared to PRESENT simulated in similar platforms. Then, Chiu C. Tan,, Haodong Wang, Sheng Zhong and Qun Li (Tan et al., 2009) developed a lightweight identity based cryptography for body sensor networks that manages security, privacy and accessibility for health care monitoring and tested it on commercially available sensors. Simulation results showed that the proposed method performs faster computation than other sensor platforms but suffered from slow query performance compared to other ciphers.

Abdul Rehman Raza, Khawir Mahmood, Muhammad Faisal Amjad, Haider Abbas (Raza et al., 2020) implemented 64, 80 and 128 bits of LED block cipher across various programming languages such as C++, Java, Python. Software efficiency and throughput was studied using 32- and 64-bit platforms using Windows and Linux operating systems. They have highlighted and studied the impact in the choice of programming language and platform on the performance of the algorithm. Results show that the choice of platform and language can affect the efficiency of an algorithm with a factor as high as 400. Finally, Norah Alassaf 1 & Adnan Gutub (Alassaf et al., 2019) have proposed further improvements in SIMON cipher that can be used to preserve medical data in an IoT setup. The work is compared with AES algorithm in terms of memory consumption and execution time. The proposed technique offers high security while maintaining a trade of between cost and performance as well as ROM and RAM memory consumption.

MOTIVATION

Health care is one of the fastest sectors to adapt to the changes made in IoT based systems. ""MarketsAndMarkets" predicts that IoHT will be worth US$ 163.2B, commercial report claims a spending of $117B, and McKinsey estimates an economic impact of more than US$ 170B" (Rodrigues et al., 2018). development of e-health systems such as electrocardiography, electroencephalogram, diabetes can be cost saving and help patients suffering with chronic diseases reduce the number of hospitals visits (Khairuddin et al., 2017). Also, the outbreak of covid-19 pandemic has created a fear in minds of people and refraining them visiting hospitals which could potentially cause them to suffer from the virus. This has enabled the IoHT sector to grow exponentially and will continue to bloom for the next few years. People are now looking for safer and less expensive ways to maintain and monitor their health. Due to the increased number of users, it has become an attractive sector for hackers.

This raises the need to develop IoT based systems with enhanced security that enables safe transfer and computation of patients' data. Security can be achieved by various methods like cryptography, block chain technology, machine learning techniques like supervised, unsupervised and reinforced learning etc. (Xiao et al., 2018). This paper focuses on simple and lightweight block ciphers that are implemented to protect the data at the sensing/physical layer of any Iot based system.

CONCERNS AND CHALLENGES IN IMPLEMENTATION OF CRYPTOGRAPHIC TECHNIQUES TO RESOURCE CONSTRAINED IoHT DEVICES

From physical sensors to computer servers, any IoT network incorporates a wide variety of platforms. This opens the door to a slew of new concerns for users, including privacy, security, compatibility, scalability, and interoperability (Nair, Tyagi, & Sreenath, 2021). IoT devices are a particularly appealing target for hackers because they interact directly with the actual environment to collect sensitive data (Feng et al., 2018). These devices can potentially be physically damaged in addition to being tapped to gather the sensitive data provided. As a result, cyber security is required, which is regarded as a key problem in the implementation of authentication, data security, availability, privacy, and accessibility (Mohd, Hayajneh, & Vasilakos, 2015). The method adopted for securing the sensitive data completely depends on the environment. The proposed method must be suitable and highly secure to the applied layer of an IoT device but should be designed in such a manner that it does not affect any of its regular activities. Conventional PC cryptographic techniques do not fit into this category as these devices are highly resource constrained.

The cryptographic technique used to preserve this information must be designed by keeping in mind the limitation of the device. The major challenges include (see Figure 2): (Singh, Sharma, Moon et al, 2017)

- Low computation power
- Lower energy
- Reduction in availability of space due to smaller size
- Reduction in memory space (ROM and RAM)
- Lower power
- Faster execution time

Figure 2. Challenges in implementation of cryptography

SOLUTIONS TO ENHANCE SECURITY IN PHYSICAL LAYER OF IoHT DEVICES

The main characteristics to be taken into consideration while choosing the right cryptographic techniques are cost, performance and security level. Performance can further be divided into subsections such as energy and power consumption, latency, computation speed, memory occupation and different attack models such as linear and differential attacks, side channel attacks and gault injection attacks (Stallings, 2017). Most of the above-mentioned issues are resolved using LWC techniques with a simple key and fewer rounds, but block cypher security is achieved by employing a rigorous internal structure such as FN, hybrid, SPN, GRX, and others to make it impervious to attacks (Thakor et al., 2021). Cryptographic techniques are categorized as symmetric and asymmetric based on the number of keys. A symmetric cryptographic technique uses the same key for encryption as well as decryption whereas asymmetric technique has two private public key pairs (Bhardwaj et al., 2017). The encryption and decryption processes of symmetric block cyphers are continuous. In a symmetric block cypher, obtaining the plain text in the reverse procedure is difficult. As a result, they outperform asymmetric cyphers since the usage of two separate keys slows down the computing process (Rashidi, 2020).

LIGHTWEIGHT CRYPTOGRAPHIC TECHNIQUES

The general architecture, encryption mechanism and the size of the plaintext and key corresponding to the LWC algorithms namely CLEFIA, KATAN and SIMON are discussed below.

CLEFIA

CLEFIA is a 128-bit symmetric block cipher which is developed based on general Feistel network (GFN) and can be implemented by using key ciphers of 128-bits, 192 bits and 256 bits. A GFN based technique is an extrapolated version of FN. The text is encrypted by spitting the word into few sub-blocks and applying FN mechanism and further performing cyclic rotation of bits based on the number of sub-blocks (Saravanan et al., 2019). It is considered as one of the best alternatives for AES, a standard encryption method adopted by the U.S government to protect sensitive data (Ertaul & Rajegowda, 2017; Jangra & Singh, 2019).

The main techniques involved are branching and number of rounds of encryption. The data processing part of CLEFIA contains 4-whitening keys, 2n number of round keys (n is number of rounds) which are 32 bits wide (see figure 3). A key scheduling process takes place in order to produce intermediate keys which are usually updated in every two rounds which in turn is expanded to derive at 2n round keys and 4 whitening keys (Shirai et al., 2007; Tezcan, 2010). The two s boxes are useful in order to face algebraic and byte ordering saturation attacks (Su et al., 2014).

The encrypted text is obtained by passing the original message through two layers of s-boxes and permutation levels.

Figure 3. CLEFIA flowchart (Saravanan et al., 2019)

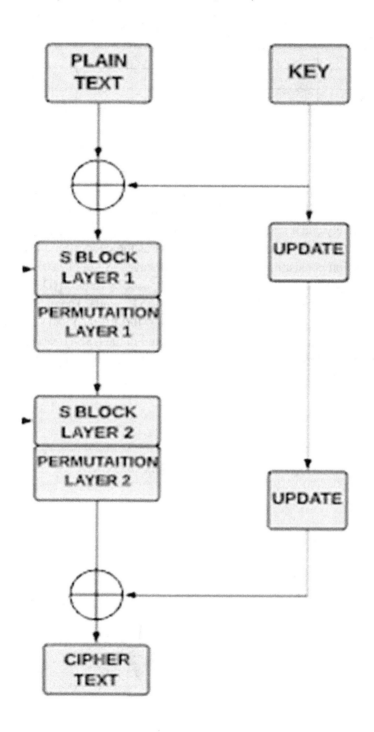

Table 1. Building block of CLEFIA algorithm

BLOCK SIZE (BITS)	KEY SIZE (BITS)	NO OF BRANCHES (D)	NO FO ROUNDS (N)
128	128	4	18
128	192	8	22
128	256	8	26

KATAN

KATAN inspired by KeeLoq, is a subsection of the block cipher family which consists of KATAN32, KATAN48 and KATAN64 with block sizes 32, 48 and 65 bits respectively and a fixed key size of 80 bis (Qatan & Damaj, 2012). It is highly notable for its simplicity and is more efficient as the encryption process runs in a parallel fashion consisting of three pipelined stages (Su et al., 2014). It follows a linear structure (LFSR) rather than NLFSR proposed by KeeLoq (De Canniere et al., 2009). The encryption process takes a total of 254 rounds. The plain text is loaded into two registers L1 and L2. In each round a few bits form L1 and L2 are computed using predefined non-linear functions and stored in the LSB (Least significant bit) of L1 and l2 respectively (Mohd, Hayajneh, & Abu Khalaf, 2015). The non-linear functions are defined as follows:

Figure 4. CLEFIA encryption structure (Mohd et al., 2018)

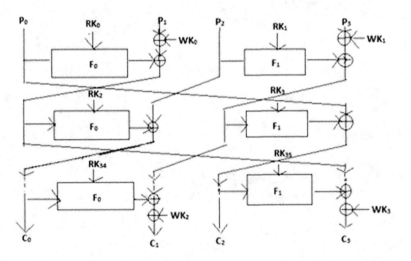

Table 2. Parameters of KATAN

PARAMETER	BLOCK SIZE (BITS)		
	32	48	64
|L1|	13	19	25
|L2|	19	29	39
{x1}, {x2}, {x3}, {x4}, {x5}	12, 7, 8, 5, 3	18, 12, 15, 7, 6	24, 15, 20, 11, 9
{y1}, {y2}, {y3}, {y4}, {y5}	18, 7, 12, 10, 8, 3	29, 19, 21, 13, 15, 6	38, 25, 33, 21, 14, 9

$$fa(L1) = L1[x1] \oplus L1[x2] \oplus (L1[x3] \cdot L1[x4]) \oplus (L1[x5] \cdot IR) \oplus keya$$
$$(1)$$

$$fb(L2) = L2[y1] \oplus L2[y2] \oplus (L2[y3] \cdot L2[y4]) \oplus (L2[y5] \cdot L2[y6]) \oplus keyb$$
$$(2)$$

Where {x} ands {y} are defined in table 2 and Keya and keyb are sub-key bits. The number of times the nonlinear function is applied in a round varied from 1 in KATAN32 to 2 and 3 in KATAN 48 and KATAN64 respectively (Abed et al., 2019).

Figure 5. KATAN Structure (Tezcan, 2010)

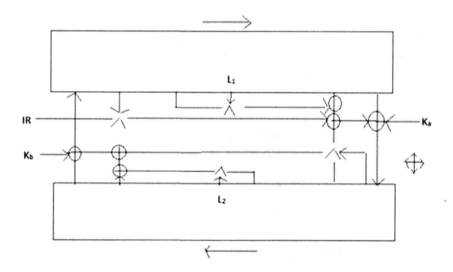

Figure 6. KATAN flowchart

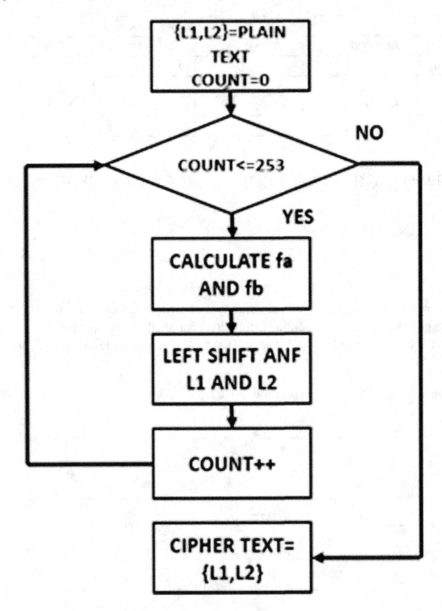

Similar to any block cipher, KATAN algorithm includes various intermediate process taking place during encryption (see Figures 5 and 6).

The IR (Irregular update) rule in KATAN ensures that not mor than seven rounds share the same pattern of updates. This ensures than not more than 7 rounds can be utilized by self-similarity attacks. Due to this, these kinds of attacks often fail in KATAN family of ciphers.

SIMON

Simon is a lightweight lock cipher proposed by National Security Agency (NSA) in 2013 as a part of Simon and Speck family of ciphers (Abed et al., 2019). The Simon cipher can be represented as Simon 2n/mn where, 2n is the block size, n is the word size and m are the number of key words. The encryption process is done using the round function which performs the following operations on the plaintext (AlKhzaimi & Lauridsen, 2013).

- BITWISE XOR
- BITWISE AND
- LEFT CIRCULAR SHIFT

The plaintext block is split into two equal parts namely left and right block. Each round performs three left shift operations of the left block then ANDs it and the resultant value is XORed with the right block and is stored in the left block (see Figures 7 and 8).

$$F(x, y, k) = (y \bigoplus ((S^1x \text{ \& } S^8x) \bigoplus S^2x) \bigoplus k) \tag{3}$$

Where, k is the key size, x and y ae the left and right blocks respectively. Simon is more suitable for hardware applications as it requires a greater number of rounds for encryption (AlAssaf et al., 2017). Due to the multiple rounds of encryption process its highly secure against integral, man in the middle and differential attacks (Manulis et al., 2016). SIMON can be implemented using variable block size (see table 3).

RESULTS AND DISCUSSION

Execution time, encryption and decryption time, memory usage during execution, and size are all used to evaluate the performance of existing cryptographic algorithms in software. These algorithms were tested using the Python programming language on a variety of platforms, including Conda, Python IDLE, and Google Colab, using an Intel i5 core CPU with a clock speed of 1.6 GHz.

Figure 7. Structure of SIMON (Abed et al., 2019)

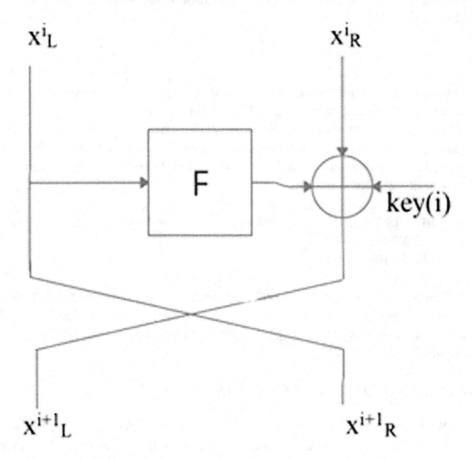

Simulation Results

The following is a sample of the results produced using CLEFIA 128-bit key encryption and decryption (see Figures 9, 10 and 11).

A tabulation of the simulated results in three platforms: Conda, Python IDLE and Google Colab is given below in table 2. The simulation is performed for variable key sizes: 128-bit, 192-bit and 256-bit with standard block size of 128-bit.

It is clear from the data that the parameters change as the platform changes. Python IDLE has the fastest algorithm, with an execution time of 4.56 seconds and encryption and decryption times of 2.285 seconds and 1.678 seconds, respectively. When compared to other platforms, Python IDLE consumes the least amount of memory. Python IDLE consumes 25 MiB of RAM, while Colab consumes 117 MiB. A total of 16.173 KB of RAM memory is used. As the size of the key increases, so does the time it takes to execute it. Because the number of rounds increases as the

Figure 8. SIMON cipher flowchart

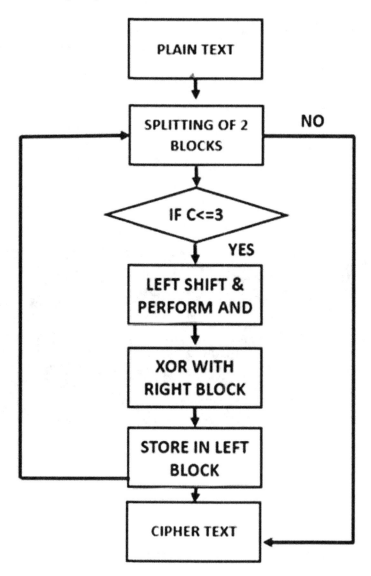

key size grows, encrypting the data takes longer. KATAN algorithm is analyzed using similar parameters used to measure the performance of CLEFIA (see Figure 12). The block size of the data is varied rather than key size whereas the opposite takes place in CLEFIA.

The above-mentioned figures show that the time of execution is really fast compared to CLEFIA.

Table 3. Parameters of SIMON

BLOCK SIZE (m bits)	KEY WORDS (n bits)	KEY SIZE (mn bits)	NUMBER OF ROUNDS
32	4	64	32
48	3	72	36
64	3	96	42
96	2	92	52
128	2	128	68
128	3	192	69
128	4	256	72

Figure 9. Performance of CLEFIA algorithm on Conda

```
-----LIGHTWEIGHT CRYPTOGRAPHY-----
'-----------CLEFIA------------
Plain text   5233100606242806050955395731361295

Encrypted text 1679830523941166824269 9517
Average elapsed time for 16-byte block 128 encryption: 2.508ms
Deccrypted text 5233100606242806050955395731361295
Average elapsed time for 16-byte block 128 decryption: 2.430ms
Total memory consumption during execution is  60.30078125 MiB


Clefia file size 16.635 KB
execution time 5.156162977218628 s
```

Figure 10. Performance of CLEFIA algorithm on Python IDLE

```
-----LIGHTWEIGHT CRYPTOGRAPHY-----
'-----------CLEFIA------------
Plain text   5233100606242806050955395731361295

Encrypted text 1679830523941166824269 9517
Average elapsed time for 16-byte block 128 encryption: 2.285ms
Deccrypted text 5233100606242806050955395731361295
Average elapsed time for 16-byte block 128 decryption: 1.678ms
Total memory consumption during execution is  25.25 MiB


Clefia file size 16.173 KB
execution time 4.4595160484313965 s
```

Figure 11. Performance of CLEFIA algorithm on Colab

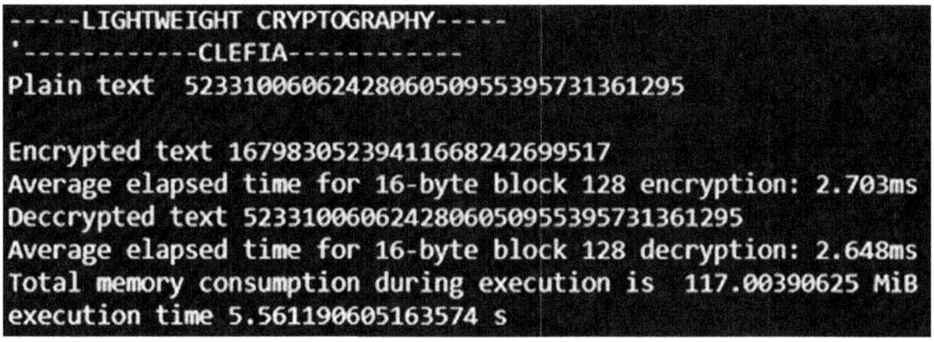

```
-----LIGHTWEIGHT CRYPTOGRAPHY-----
*------------CLEFIA------------
Plain text  5233100606242280605095535731361295

Encrypted text 1679830523941166824269517
Average elapsed time for 16-byte block 128 encryption: 2.703ms
Deccrypted text 5233100606242280605095535731361295
Average elapsed time for 16-byte block 128 decryption: 2.648ms
Total memory consumption during execution is  117.00390625 MiB
execution time 5.561190605163574 s
```

Table 4. Comparison of CLEFIA on different platforms

PLATFORM	BLOCK SIZE	KEY SIZE	NO. OF ROUNDS	EXECUTION TIME (s)	ENCRYPTION TIME m(s)	DECRYPTION TIME (ms)	MEMORY OCCUPIED DURING EXECUTION (MiB)
CONDA	128	128	18	5.159	2.508	2.430	59.45
		192	22	7.061	3.636	3.218	
		256	26	7.938	3.655	4.079	
PYTHON IDLE	128	128	18	4.46	2.285	1.678	25.29
		192	22	5.293	2.638	2.250	
		256	26	5.952	3.022	2.453	
GOOGLE COLAB	128	128	18	5.56	2.703	2.648	117.98
		192	22	7.486	3.634	3.640	
		256	26	8.275	4.024	4.040	

Figure 12. Performance of KATAN algorithm in Conda

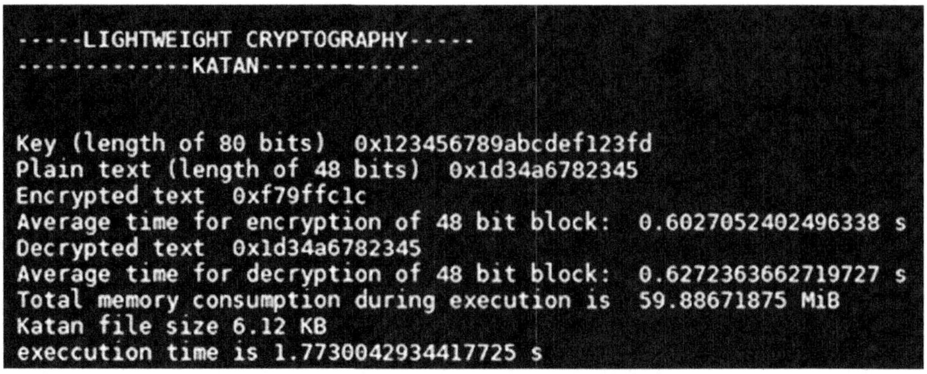

```
-----LIGHTWEIGHT CRYPTOGRAPHY-----
------------KATAN------------

Key (length of 80 bits)  0x123456789abcdef123fd
Plain text (length of 48 bits)  0x1d34a6782345
Encrypted text  0xf79ffc1c
Average time for encryption of 48 bit block:  0.6027052402496338 s
Decrypted text  0x1d34a6782345
Average time for decryption of 48 bit block:  0.6272363662719727 s
Total memory consumption during execution is  59.88671875 MiB
Katan file size 6.12 KB
execcution time is 1.7730042934417725 s
```

Figure 13. Performance of KATAN algorithm in Python IDLE

```
-----LIGHTWEIGHT CRYPTOGRAPHY-----
--------------KATAN------------

Key (length of 80 bits)  0x123456789abcdef123fd
Plain text (length of 48 bits)  0x1d34a6782345
Encrypted text  0xf79ffc1c
Average time for encryption of 48 bit block:  0.8816704750061035 s
Decrypted text  0x1d34a6782345
Average time for decryption of 48 bit block:  0.6103410720825195 s
Total memory consumption during execution is  25.15234375 MiB
Katan file size 6.12 KB
execcution time is 2.1995322704315186 s
```

Figure 14. Performance of KATAN algorithm in Colab

```
-----LIGHTWEIGHT CRYPTOGRAPHY-----
--------------KATAN------------

Key (length of 80 bits)  0x123456789abcdef123fd
Plain text (length of 48 bits)  0x1d34a6782345
Encrypted text  0xf79ffc1c
Average time for encryption of 48 bit block:  0.8102800846099854 s
Decrypted text  0x1d34a6782345
Average time for decryption of 48 bit block:  0.8175473213195801 s
Total memory consumption during execution is  117.14453125 MiB
execcution time is 2.1356799602508545 s
```

Table 5. Comparison of KATAN on different platforms

PLATFORM	WORD SIZE	KEY SIZE	EXECUTION TIME (s)	ENCRYPTION TIME (s)	DECRYPTION TIME (s)	MEMORY OCCUPIED DURING EXECUTION (MiB)
CONDA	32 48 64	80	1.763 1.771 1.835	0.632 0.619 0.654	0.628 0.616 0.665	55.32
PYTHON IDLE	32 48 64	80	2.268 2.32 2.386	1.039 0.946 0.977	0.482 0.535 0.569	25.254
GOOGLE COLAB	32 48 64	80	2.128 2.153 2.132	0.809 0.799 0.805	0.810 0.846 0.816	118.46

Figure 15. Performance of SIMON in Conda

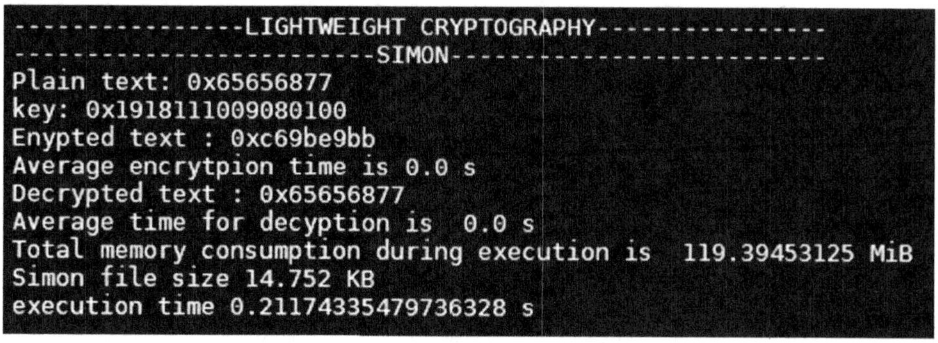

```
---------------LIGHTWEIGHT CRYPTOGRAPHY----------------
-----------------------SIMON-------------------------
Plain text: 0x65656877
key: 0x1918111009080100
Enypted text : 0xc69be9bb
Average encrytpion time is 0.0 s
Decrypted text : 0x65656877
Average time for decyption is  0.0 s
Total memory consumption during execution is  119.39453125 MiB
Simon file size 14.752 KB
execution time 0.21174335479736328 s
```

Figure 16. Performance of SIMON in Python IDLE

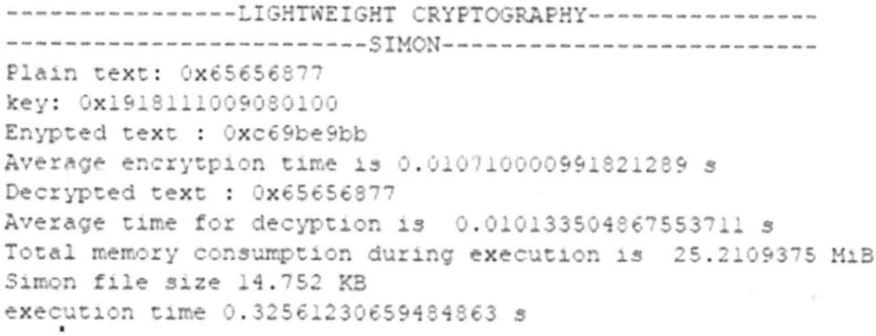

```
---------------LIGHTWEIGHT CRYPTOGRAPHY----------------
-----------------------SIMON-------------------------
Plain text: 0x65656877
key: 0x1918111009080100
Enypted text : 0xc69be9bb
Average encrytpion time is 0.010710000991821289 s
Decrypted text : 0x65656877
Average time for decyption is  0.010133504867553711 s
Total memory consumption during execution is  25.2109375 MiB
Simon file size 14.752 KB
execution time 0.32561230659484863 s
```

Table 5 explains the simulation results obtained by varying the platform and block size of the plain text. The block size is varied from 32-bits to 48 and 64 bits.

Figure 17. Performance of SIMON in Colab

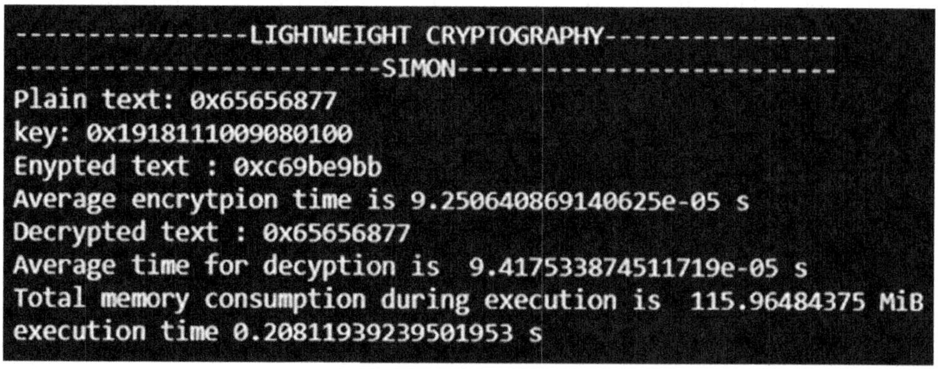

```
---------------LIGHTWEIGHT CRYPTOGRAPHY----------------
-----------------------SIMON-------------------------
Plain text: 0x65656877
key: 0x1918111009080100
Enypted text : 0xc69be9bb
Average encrytpion time is 9.250640869140625e-05 s
Decrypted text : 0x65656877
Average time for decyption is  9.417533874511719e-05 s
Total memory consumption during execution is  115.96484375 MiB
execution time 0.20811939239501953 s
```

Table 6. Comparison of SIMON on different platforms

PLATFORM	BLOCK SIZE	KEY SIZE	EXECUTION TIME (s)	ENCYRPTION TIME (ms)	DECYRTOPIN TIME (ms)	MEMORY OCCUPIED DURING EXECUTION (MiB)
PYHTON IDLE	32	64	0.326	9.85	8.69	25.211
	48	72	0.339	10.12	9.98	25.17
	64	96	0.318	10.71	10.13	25.18
	128	128	0.323	11.356	9.569	25.141
	128	192	0.347	13.33	10.59	25.24
	128	256	0.331	11.67	11.39	25.15
CONDA	32	64	0.212	Failed	Failed	54.32
	48	72	0.227	Failed	Failed	55.76
	64	96	0.231	9.95	Failed	55.86
	128	128	0.221	Failed	7.43	55.82
	128	192	0.223	Failed	Failed	55.68
	128	256	0.215	Failed	Failed	56.016
GOOGLE COLAB	32	64	0.208	0.0925	0.0941	115.96
	48	72	0.209	0.0954	0.107	115.29
	64	96	0.212	0.0942	0.159	116.97
	128	128	0.208	0.133	0.156	116.44
	128	192	0.209	1.39	0.346	117.16
	128	256	0.210	2.17	0.938	117.17

In contrary to the results observed in CLEFIA, KATAN performs better in Conda environment compared to Python IDLE and Google Colab. The memory occupation during execution is slightly higher than CLEFIA. KATAN algorithm executes faster with a time of just 1.763 seconds and requires only 0.6 seconds approximately to encrypt and decrypt the data. Similar results are obtained for SIMON cipher (see Figure 15, 16 and 17).

It can be observed from the figure that SIMON cipher performs better than CLEFIA but not KATAN in term of execution time.

Table 6 illustrates that the simulation results obtained for SIMON cipher does not flow a particular trend and not all simulations were successful. However, the simulations executed using Python IDLE and Colab are quite successful. The encryption and decryption times vary almost linearly with the key and block sizes.

Performance Analysis

The performance of the techniques is analyzed using the following parameters: file size, execution time, encryption and decryption times and number of rounds. A comparison of the simulated cryptographic techniques on the basis of number of rounds with key sizes and block sizes is studied (see Figure 18). The strength

Figure 18. Number of rounds of encryption

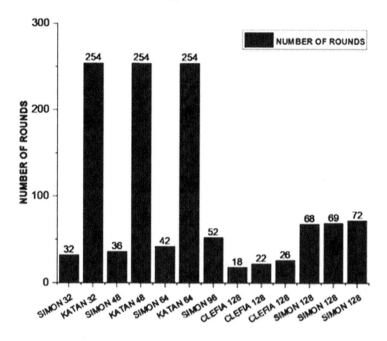

Figure 19. File size

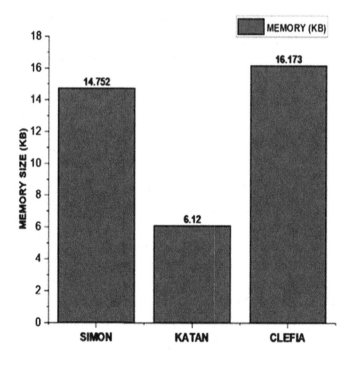

Figure 20. Execution time

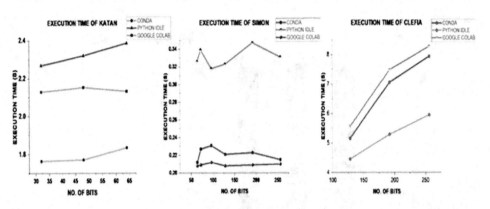

of a cryptographic technique increases with the increase in number of encryptions rounds as it becomes difficult to attack and go through multiple layers According to thumb rule the security of any block cipher is half the key size provided there are no possible methods better than brute force to crack the cipher (Viega, 2003). The future research in hardware and software is motivated by least memory consumption as IoT devices are highly memory constrained. The efficiency of any LWC algorithm highly depends on the memory occupation.

KATAN is more secure than CLEFIA and SIMON due to a greater number of encryption rounds. KATAN is also more compact than the other techniques as it occupies the least memory space of 6.12 Kbytes and surpasses SIMON and CLEFIA by 58.5% and 62.15% respectively (see Fig 18 and 19). According to the graphs, KATAN is more suitable to be applied in IoHT devices due to greater security and lesser memory occupation followed by SIMON and CLEFIA. Another important performance metric is the execution time. The total execution time includes the time taken for encryption, decryption and generation of round keys. Individual performance of each algorithm is tested in various platforms. Comparison of the execution times of the KATAN, CLEFIA and SIMON tested on various platforms is studied (see Fig. 14).

SIMON appears to be the fastest, followed by KATAN and CLEFIA, according to the graph (see Figure 18). SIMON is approximately 88% and 96% faster than KATAN and CLEFIA respectively. The execution time varies with platform and method used in the LWC algorithms. SIMON performs best in Google Colab whereas KATAN and CLEFIA perform better in Conda and IDLE respectively. The execution speed of the algorithm depends on the number of keys generated and the number of rounds. SIMON uses simple XOR and AND logical operations and is implemented in text sizes of 32,48,64 and 128 bits with minimum key size and number of rounds

Figure 21. Average encryption time

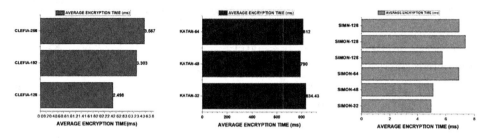

of encryption (Gennaro et al., 2015). This simple yet secure implementation makes SIMON the fastest algorithm compared to other techniques. Another parameter of relevance is the encryption time of various LWC algorithms. (Figure 21)

From the graph we can say that SIMON again demands the least encryption time with a minor difference from CLEFIA. Even though CLEFIA has lesser rounds of encryption compared to SIMON and KATAN, the multiple key scheduling processes and permutation levels takes more time than encryption of data using simple logical expressions and shifting operations. Whereas on the other hand, KATAN is very slow compared to other techniques as it goes through 254 algorithm rounds for all block sizes (Tao et al., 2019).

FUTURE SCOPE

IoT applications are growing rapidly day by day and as most of the industries are moving towards IoT, energy consumption is one of the main constraints of the IoT world (Dhanda et al., 2020; Singh, Sharma, & Moon, 2017). The limited computational capabilities and resource constraints make it a vulnerable target for hackers (Alladi et al., 2020). IoHT is a field that is widely in use now. It deals with millions of patients' health information which needs to be secured in order to prevent misuse. In future, light weight cryptographic encryption in IoHT can improve the security level of the system and help to make the devices more efficient and secure.

Lightweight block ciphers are efficient in both hardware as well as software. Asymmetric block ciphers, stream, hash and elliptical curve functions are other available techniques which have a high potential to be employed in these devices to secure patients' information in IoHT (Chen et al., 2017). The future work includes study of software performance using other programming languages such as C++, Java and other operating systems such as Linux. Further the hardware performance can be studied by implementing these techniques in real time embedded systems, ARM-based microprocessors and dedicated integrated circuits which are widely used in IoHT industry to observe various parameters like circuit-footprint, throughput, latency, energy and power consumptions in order to design ultra-low power IoT devices.

In the last, in (Gudeti et al., 2020; Tyagi et al., 2022; Pal et al., 2022; Nair, Kumari, Tyagi et al, 2021; Nair & Tyagi, 2021; Pramod, 2020; Rekha et al., 2020; Rekha et al., 2019; Tyagi, 2021) authors have recommended others to read these research efforts, to know more information about IoHT and role of AI, Computer Vision, or Machine learning techniques respect to this sensitive area/ useful application. We hope that readers/ researchers will find suitable problem for themselves/ for their research work (from these research works).

CONCLUSION

This paper analyses the software implementation of some widely used symmetric block ciphers in IoHT devices namely CLEFIA, KATAN and SIMON. The performance of these algorithms in terms of speed, and memory consumption was studied platforms like Python IDLE, Colab and Conda. The simulation results showed that KATAN occupies less memory than CLEFIA and SIMON by 58.5% and 62.15% and a greater number of rounds for more secure encryption. Whereas SIMON proves to be better in terms of speed by 88% and 96% than KATAN and CLEFIA due to a simpler yet secure method of encryption. As a result, it can be stated that KATAN is more suitable for devices with high resource constraints in terms of memory and SIMON is preferable for solutions which can respond to sensory inputs within specified times like real-time embedded systems.

ACKNOWLEDGMENT

The authors want to thank Centre for Advanced Data Science and School of Computer Science and Engineering, Vellore Institute of Technology, Chennai for providing their kind support to complete this research work on time.

REFERENCES

Abed, S., Jaffal, R., Mohd, B. J., & Alshayeji, M. (2019). FPGA Modeling and Optimization of a SIMON Lightweight Block Cipher. *Sensors (Basel)*, *19*(4), 913. doi:10.339019040913 PMID:30795605

Aboushosha, B., Ramadan, R. A., Dwivedi, A. D., El-Sayed, A., & Dessouky, M. M. (2020). SLIM: A Lightweight Block Cipher for Internet of Health Things. *IEEE Access: Practical Innovations, Open Solutions*, 8, 203747–203757. doi:10.1109/ACCESS.2020.3036589

AlAssaf, N., AlKazemi, B., & Gutub, A. (2017). Applicable Light-Weight Cryptography to Secure Medical Data in IoT Systems. *Journal of Research in Engineering and Applied Sciences*, 2(2), 50–58. doi:10.46565/jreas.2017.v02i02.002

Alassaf, N., Gutub, A., Parah, S. A., & Al Ghamdi, M. (2019). Enhancing speed of SIMON: A light-weight-cryptographic algorithm for IoT applications. *Multimedia Tools and Applications*, 78(23), 32633–32657. doi:10.100711042-018-6801-z

AlKhzaimi, H., & Lauridsen, M. M. (2013). Cryptanalysis of the SIMON Family of Block Ciphers. *IACR Cryptol. ePrint Arch.*, 543.

Alladi, T., Chamola, V., Sikdar, B., & Choo, K.R. (2020). Consumer IoT: Security Vulnerability Case Studies and Solutions. *IEEE Consum. Electron.*, 17–25.

Bhardwaj, I., Kumar, A., & Bansal, M. (2017). A review on lightweight cryptography algorithms for data security and authentication in IoTs. *Proc. 4th Int. Conf. Signal Process., Comput. Control (ISPCC)*, 504–509. 10.1109/ISPCC.2017.8269731

Biryukov, A., & Perrin, L. P. (2017). *State of the art in lightweight symmetric cryptography*. Univ. Luxembourg Library, Esch-sur-Alzette, Luxembourg, Tech. Rep. 10993/31319.

Biswas, K., Muthukkumarasamy, V., & Singh, K. (2015). An encryption scheme using chaotic map and genetic operations for wireless sensor networks. *IEEE Sensors Journal*, 15(5), 2801–2809. doi:10.1109/JSEN.2014.2380816

Butpheng, Yeh, & Xiong. (2020). Security and Privacy in IoT-Cloud-Based e-Health Systems—A Comprehensive Review. *Symmetry, 12*(7).

Chaudhury, S., Paul, D., & Mukherjee, R. (2017). Haldar S Internet of Thing based healthcare monitoring system. In *Industrial automation and electromechanical engineering conference (IEMECON), 2017 8th annual* (pp. 346–349). IEEE. doi:10.1109/IEMECON.2017.8079620

Chen, L., Thombre, S., Jarvinen, K., Lohan, E. S., Alen-Savikko, A., Leppakoski, H., Bhuiyan, M. Z. H., Bu-Pasha, S., Ferrara, G. N., Honkala, S., Lindqvist, J., Ruotsalainen, L., Korpisaari, P., & Kuusniemi, H. (2017). Robustness, Security and Privacy in Location-Based Services for Future IoT: A Survey. *IEEE Access: Practical Innovations, Open Solutions*, 5, 8956–8977. doi:10.1109/ACCESS.2017.2695525

De Canniere, C., Dunkelman, O., & Knežević, M. (2009). KATAN and KTANTAN—A family of small and efficient hardware-oriented block ciphers. In *Proc. Int. Workshop Cryptograph. Hardw. Embedded Syst.* Springer. 10.1007/978-3-642-04138-9_20

Dhanda, S. S., Singh, B., & Jindal, P. (2020). Lightweight Cryptography: A Solution to Secure IoT. *Wireless Personal Communications*, *112*(3), 1947–1980. doi:10.100711277-020-07134-3

El-hajj, M., Chamoun, M., Fadlallah, A., & Serhrouchni, A. (2017). Analysis of authentication techniques in internet of things (IoT). In *Cyber security in networking conference (CSNet), 1st* (pp. 1–3). IEEE. doi:10.1109/CSNET.2017.8242006

Engineering, A. A. B. (2017). *Internet of Medical Things Revolutionizing Healthcare.* https://aabme.asme.org/posts/internet-of-medical-thingsrevolutionizing-healthcare

Ertaul, L., & Rajegowda, S. K. (2017). Performance Analysis of CLEFIA, PICCOLO, TWINE Lightweight Block Ciphers in IoT Environment. In *Proceedings of the International Conference on Security and Management (SAM). The Steering Committee of The World Congress in Computer Science, Computer Engineering and Applied Computing (WorldComp)* (pp. 25-31). Academic Press.

Fan, X., Mandal, K., & Gong, G. (2013). Wg-8: a lightweight stream cipher for resource-constrained smart devices. *Proceeding of International Conference on Heterogeneous Networking for Quality, Reliability.* 10.1007/978-3-642-37949-9_54

Feng, W., Qin, Y., Zhao, S., & Feng, D. (2018). AAoT: Lightweight attestation and authentication of low-resource things in IoT and CPS. *Computer Networks*, *134*, 167–182. doi:10.1016/j.comnet.2018.01.039

Fuzon. (2019). *Internet of Medical Things (IoMT): New Era in Healthcare Industry.* Academic Press.

Gubbi, J., Buyya, R., Marusic, S., & Palaniswami, M. (2013). Internet of things (iot): A vision, architectural elements, and future directions. *Future Generation Computer Systems*, *29*(7), 1645–1660. doi:10.1016/j.future.2013.01.010

Gudeti, B., Mishra, S., Malik, S., Fernandez, T. F., Tyagi, A. K., & Kumari, S. (2020). A Novel Approach to Predict Chronic Kidney Disease using Machine Learning Algorithms. *2020 4th International Conference on Electronics, Communication and Aerospace Technology (ICECA)*, 1630-1635. 10.1109/ICECA49313.2020.9297392

Harikrishnan, T., & Babu, C. (2015). Cryptanalysis of hummingbird algorithm with improved security and throughput. *2015 International Conference on VLSI Systems, Architecture, Technology and Applications (VLSI-SATA)*, 1-6. 10.1109/VLSI-SATA.2015.7050460

Ismail, L., Materwala, H., & Zeadally, S. (2019). Lightweight Blockchain for Healthcare. *IEEE Access: Practical Innovations, Open Solutions*, 7, 149935–149951. doi:10.1109/ACCESS.2019.2947613

Jangra, M., & Singh, B. (2019). Performance analysis of CLEFIA and PRESENT lightweight block ciphers. *Journal of Discrete Mathematical Sciences and Cryptography*, 22(8), 1489–1499. doi:10.1080/09720529.2019.1695900

Khairuddin, A. M., Azir, K. N. F. K., & Kan, P. E. (2017). Limitations and future of electrocardiography devices: A review and the perspective from the Internet of Things. *International Conference on Research and Innovation in Information Systems*, 1-7. 10.1109/ICRIIS.2017.8002506

Kölbl, S., Leander, G., & Tiessen, T. (2015). Observations on the SIMON Block Cipher Family. In R. Gennaro & M. Robshaw (Eds.), Lecture Notes in Computer Science: Vol. 9215. *Advances in Cryptology — CRYPTO 2015. CRYPTO 2015.* doi:10.1007/978-3-662-47989-6_8

Kondo, K., Sasaki, Y., & Iwata, T. (2016). On the Design Rationale of Simon Block Cipher: Integral Attacks and Impossible Differential Attacks against Simon Variants. In M. Manulis, A. R. Sadeghi, & S. Schneider (Eds.), Lecture Notes in Computer Science: Vol. 9696. *Applied Cryptography and Network Security. ACNS 2016.* doi:10.1007/978-3-319-39555-5_28

Li, L., Liu, B., & Wang, H. (2016). QTL: A new ultra-lightweight block cipher. *Microprocessors and Microsystems*, 45, 45–55. doi:10.1016/j.micpro.2016.03.011

Madhav, A. V. S., & Tyagi, A. K. (2022). The World with Future Technologies (Post-COVID-19): Open Issues, Challenges, and the Road Ahead. In A. K. Tyagi, A. Abraham, & A. Kaklauskas (Eds.), *Intelligent Interactive Multimedia Systems for e-Healthcare Applications*. Springer. doi:10.1007/978-981-16-6542-4_22

McKay, K., Bassham, L., Turan, M. S., & Mouha, N. (2017). *Report on Lightweight Cryptography (Nistir8114)*. NIST.

Mishra, S., & Tyagi, A. K. (2022). The Role of Machine Learning Techniques in Internet of Things-Based Cloud Applications. In S. Pal, D. De, & R. Buyya (Eds.), *Artificial Intelligence-based Internet of Things Systems. Internet of Things (Technology, Communications and Computing)*. Springer. doi:10.1007/978-3-030-87059-1_4

Mohd, B. J., Hayajneh, T., & Abu Khalaf, Z. (2015). Optimization and modeling of FPGA implementation of the KATAN Cipher. *2015 6th International Conference on Information and Communication Systems (ICICS)*, 68-72. 10.1109/IACS.2015.7103204

Mohd, B. J., Hayajneh, T., Ahmad Yousef, K. M., Khalaf, Z. A., & Bhuiyan, M. Z. A. (2018). Hardware design and modeling of lightweight block ciphers for secure communications. *Future Generation Computer Systems*, *83*, 83. doi:10.1016/j.future.2017.03.025

Mohd, B. J., Hayajneh, T., & Vasilakos, A. V. (2015). A survey on lightweight block ciphers for low-resource devices: Comparative study and open issues. *Journal of Network and Computer Applications*, *58*, 73–93. doi:10.1016/j.jnca.2015.09.001

Nair, M. M., Kumari, S., Tyagi, A. K., & Sravanthi, K. (2021). Deep Learning for Medical Image Recognition: Open Issues and a Way to Forward. In *Proceedings of the Second International Conference on Information Management and Machine Intelligence. Lecture Notes in Networks and Systems* (vol. 166). Springer.

Nair, M. M., & Tyagi, A. K. (2021). Privacy: History, Statistics, Policy, Laws, Preservation and Threat Analysis. Journal of Information Assurance & Security, 16(1), 24-34.

Nair, M. M., Tyagi, A. K., & Sreenath, N. (2021). The Future with Industry 4.0 at the Core of Society 5.0: Open Issues, Future Opportunities and Challenges. *2021 International Conference on Computer Communication and Informatics (ICCCI)*, 1-7. 10.1109/ICCCI50826.2021.9402498

Pramod, A. (2020). *Machine Learning and Deep Learning: Open Issues and Future Research Directions for Next Ten Years. In Computational Analysis and Understanding of Deep Learning for Medical Care: Principles, Methods, and Applications, 2020*. Wiley Scrivener.

Qatan, F. M., & Damaj, I. W. (2012). High-speed KATAN ciphers on-a-chip. *2012 International Conference on Computer Systems and Industrial Informatics*, 1-6.

Rashidi, B. (2020). Efficient and flexible hardware structures of the 128-bit CLEFIA block cipher. *IET Computers & Digital Techniques*, *14*(2), 69–79. doi:10.1049/iet-cdt.2019.0157

Raza, A. R., Mahmood, K., Amjad, M. F., Abbas, H., & Afzal, M. (2020). On the efficiency of software implementations of lightweight block ciphers from the perspective of programming languages. *Future Generation Computer Systems*, *104*, 43–59. doi:10.1016/j.future.2019.09.058

Rekha, Reddy, & Tyagi. (2020). KDOS - Kernel Density based Over Sampling - A Solution to Skewed Class Distribution. *Journal of Information Assurance and Security, 15*(2), 44-52.

Rekha, G., Tyagi, A. K., & Krishna Reddy, V. (2019). Solving Class Imbalance Problem Using Bagging, Boosting Techniques, with and without Noise Filter Method. *International Journal of Hybrid Intelligent Systems*, *15*(2), 67–76. doi:10.3233/HIS-190261

Rodrigues, J. J. P. C., De Rezende Segundo, D. B., Junqueira, H. A., Sabino, M. H., Prince, R. M., Al-Muhtadi, J., & De Albuquerque, V. II. C. (2018). Enabling Technologies for the Internet of Health Things. *IEEE Access: Practical Innovations, Open Solutions*, *6*, 13129–13141. doi:10.1109/ACCESS.2017.2789329

Saravanan, P., Rani, S. S., Rekha, S. S., & Jatana, H. S. (2019). An Efficient ASIC Implementation of CLEFIA Encryption/Decryption Algorithm with Novel S-Box Architectures. *2019 IEEE 1st International Conference on Energy, Systems and Information Processing (ICESIP)*, 1-6. 10.1109/ICESIP46348.2019.8938329

Shirai, T., Shibutani, K., Akishita, T., Moriai, S., & Iwata, T. (2007, January). The 128-bit blockcipher CLEFIA. In *Fast software encryption* (pp. 181–195). Springer Berlin Heidelberg. doi:10.1007/978-3-540-74619-5_12

Singh, S., Sharma, P. K., & Moon, S. Y. (2017). *Advanced lightweight encryption algorithms for IoT devices: survey, challenges and solutions. J Ambient Intell Human Comput.*

Singh, S., Sharma, P. K., Moon, S. Y., & Park, J. H. (2017). Advanced lightweight encryption algorithms for IoT devices: Survey, challenges and solutions. *Journal of Ambient Intelligence and Humanized Computing*, *4*, 1–18. doi:10.100712652-017-0494-4

Stallings. (2017). *Cryptography and Network Security: Principles and Practice.* Academic Press.

Su, S., Dong, H., Fu, G., Zhang, C., & Zhang, M. (2014). A White-Box CLEFIA implementation for mobile devices. *2014 Communications Security Conference (CSC 2014)*, 1-8.

Tan, C. C., Wang, H., Zhong, S., & Li, Q. (2009, November). IBE-Lite: A Lightweight Identity-Based Cryptography for Body Sensor Networks. *IEEE Transactions on Information Technology in Biomedicine, 13*(6), 926–932. doi:10.1109/TITB.2009.2033055 PMID:19789117

Tao, H., Bhuiyan, M. Z. A., Abdalla, A. N., Hassan, M. M., Zain, J. M., & Hayajneh, T. (2018). Secured data collection with hardware-based ciphers for IoT-based healthcare. *IEEE Internet of Things Journal*, 1–10.

Tao, H., Bhuiyan, M. Z. A., Abdalla, A. N., Hassan, M. M., Zain, J. M., & Hayajneh, T. (2019). Secured Data Collection with Hardware-Based Ciphers for IoT-Based Healthcare. *IEEE Internet of Things Journal, 6*(1), 410–420. doi:10.1109/JIOT.2018.2854714

Tawalbeh, Muheidat, Tawalbeh, & Quwaider. (2020). IoT Privacy and Security: Challenges and Solutions. *Applied Sciences, 10*(12).

Tezcan, C. (2010). The improbable differential attack: Cryptanalysis of reduced round CLEFIA. In *Progress in Cryptology-INDOCRYPT 2010* (pp. 197-209). Springer Berlin Heidelberg.

Thakor, V. A., Razzaque, M. A., & Khandaker, M. R. A. (2021). Lightweight Cryptography Algorithms for Resource-Constrained IoT Devices: A Review, Comparison and Research Opportunities. *IEEE Access: Practical Innovations, Open Solutions, 9*, 28177–28193. doi:10.1109/ACCESS.2021.3052867

Toshihiko, O. (2017). Lightweight cryptography applicable to various IoT devices. *NEC Tech. J., 12*(1), 67–71.

Tyagi. (2021, October). AARIN: Affordable, Accurate, Reliable and INnovative Mechanism to Protect a Medical Cyber-Physical System using Blockchain Technology. *IJIN, 2*, 175–183.

Ullah, A., Sehr, I., Akbar, M., & Ning, H. (2018). *FoG assisted secure De-duplicated data dissemination in smart healthcare IoT. In 2018 IEEE international conference on smart internet of things (SmartIoT)*. IEEE. doi:10.1109/SmartIoT.2018.00038

Viega, M. M. J. (2003). *Secure Programming Cookbook for C and C++: Recipes for Cryptography, Authentication, Input Validation & More*. Academic Press.

Xiao, L., Wan, X., Lu, X., Zhang, Y., & Wu, D. (2018). IoT Security Techniques Based on Machine Learning: How Do IoT Devices Use AI to Enhance Security? IEEE Signal Processing Magazine, 35(5), 41-49.

Chapter 10
Edge Computing Enabled by 5G for Computing Offloading in the Industrial Internet of Things

Vinodhini Mani
Sathyabama Institute of Science and Technology, India

Kavitha C.
Sathyabama Institute of Science and Technology, India

Baby Shamini P.
https://orcid.org/0000-0003-0488-9145
RMK Engineering College, India

S. R. Srividhya
Sathyabama Institute of Science and Technology, India

ABSTRACT

Game theory and 5G data transferring structures are used to enhance data offloading of computational data. All such solutions not only reduce the load on the cloud by processing edge data and information but also play an important role in privacy and security by ensuring which data communication is locally translated into a network that directly connects the user equipment and then sends the local server to the network core of the organization. With a combination of 5G structure and game theory, we can comfortably handle data transmission, data security, and also system efficiency. The game theory approach with the dynamic computation offloading algorithm (DCOA) allows IIoT (industry internet of things) devices to make decisions spontaneously and reduce computational offloading while transferring data to the local server. The small range of 5G makes it possible to use a small cell network run by a central hub. As a result of its higher frequency, 5G follows a different wireless spectrum structure (N-RAM) and is enhanced by slicing the network.

DOI: 10.4018/978-1-6684-3921-0.ch010

INTRODUCTION

The Industrial Internet of Things (IIoT) combines a broad range of diverse, disparate, and advanced technological solutions for money-associated objectives, technological upgrades, and reduced human limitations, to exchange information (e.g., data sharing, requirement sharing, and data swapping). Most frequently, the networks are connected to central servers and over 4G networks. The most popular technology is currently 4G, but it faces serious adaptability problems and has little activity. Constant changes in network focus make 5G the ideal way to tackle the flexibility issues of businesses. Business flexibility is best addressed with 5G progress. It does not appear that 5G will be set up for traditional game plans yet, but associations are starting to incorporate different items and entities that can be applied to it. The transition will be a multi-year adventure since establishing partnerships requires secure points of interest, age consistency, and solid circumstances to be formed before the whole opening. In the far-flung future, 5 G innovation in assembly, transport, production, and mining zones would contribute over \$5+ billion (F. Bonomi., et.al 2012). IoT shift, for example, is largely enabled by accessibility, but vertical markets digitize their businesses as well through defined cloud and edge processing organizations with a view to enhancing capacities, redesigning the costs of production, and enhancing protection, which offers the best advantages. The Industrial Internet of Things (IIoT) combines a broad range of diverse, disparate, and advanced technological solutions for money-associated objectives, technological upgrades, and reduced human limitations, to exchange information (e.g., data sharing, requirement sharing, and data swapping). Most frequently, they are connected via a central server and a 4G network. As of now, 4G is the most widely used technology, but it has serious adaptability problems and is largely inactive. The constant changes in network focus make 5G the ideal way to tackle the flexibility issues in businesses. Business flexibility is best addressed with 5G progress. It does not appear that 5G will be set up for traditional game plans yet, but associations are starting to incorporate different items and entities that can be applied to it. The transition will be a multi-year adventure since establishing partnerships requires secure points of interest, age consistency, and solid circumstances to be formed before the whole opening. In the far-flung future, 5 G innovation in assembly, transport, production, and mining zones would contribute over \$5+ billion (M. Chen.,et.al 2015). The Internet of associated vehicles is utilized to gather ongoing traffic citation conditions for transportation control frameworks, and the figuring errands are accessible to be resource offloaded from the vehicles to the edge processing gadgets (ECDs) for usage. In spite of various advantages of IoV and ECDs, the remote correspondence for calculation offloading builds the danger of security spillage, which may therefore prompt following, character altering, and virtual vehicle capturing. Accordingly, it stays a test to keep away from security clashes for calculation offloading to the ECDs in IoV. To address this test, an edge registering empowered calculation offloading strategy, named ECO, with protection conservation for IoV is proposed in this paper. In fact, the security clashes with processing errands in IoV are broken down in a formalized way. In regards to (Wang et al., 2019), Edge Computing (MEC) is Prospective to meet and support the ever-increasing requirements of postponement delicate and vital in 5G. With the development of 5G-enabled traffic, the board framework is currently being tested, in part due to its ability to achieve ultra-low idleness and universal availability. Additionally, edge hubs are constrained in terms of processing assets and limiting their capacity, and for this reason, Computation offloading is a critical issue. This paper advances a crossbreed computation offloading structure for ongoing traffic on the board in 5G. In particular, they consider both Non-Orthogonal Multi-Access empowered and VTV unloading derived traffic. The research issue is figured to be a joint undertaking dispersion, sub-channel tasks, and forced assignment issues, with the goal of expanding the overall offloading rate. From that point forward, they demonstrate its PN-hardness and partition into multiple discussion problems, which is to be. Execution assessments outline adequacy structures. (Xu et al., 2019) Edge registration rises as novel figuring worldview to offload processing undertakings from client supplies (UIEs) latest generation boundary nodes systems, which unquestionably breaks confinement of UIEs in a specific way. (Craciunescu et al., 2020) In this paper, they give a novel application mindful of client affiliation and assets allotment structure, i.e., AURA-5G, which uses a joint advancement procedure to achieve the equivalent. Solidly, our procedure considers all the genuine system requirements that will be common in the 5G arrangement just like common sense organizational situations. Also, AURA-5G, being an application mindful system, consider the asset prerequisites of both eMBB and mMTC administrations while playing out the enhancement task. (Li et al., 2021) With the development of IoT, applications Cloud design ends up being wasteful in taking care of enormous measures of information, for the most part as a result of variable inertness and constrained data transmission. Progressively explicit, significant necessities of Industrial Internet of Things (IIoT) as a controlled and ongoing dynamic can't be tended to. These restrictions alongside expanding knowledge in lower levels of information transmission engineering prompted the advancement of the halfway edge preparation layer, closer to the procedure, empowering dispersed figures and close to ongoing correspondence. The dynamic computation offloading algorithm (DCOA) we propose is based on stochastic optimization. DCOA is a distributed offloading algorithm that dynamically responds to changing environments through distributed computation offloading. Computational offloading is a rising zone of research that includes dividing and offloading asset-concentrated pieces of an application onto the Cloud. It is a piece of another processing worldview known as versatile distributed computing. Computational offloading is an arrangement to run asset-escalated applications on a mobile phone without agreeing on execution time or the extent to which gadget-end assets are exploited. There are different models to accomplish calculation offloading (technique, photo, and highlight offloading) but they all point to comparative goals of growth in execution and vitality sparing. With the expansion of device vulnerabilities and cloud finishes, the security and safety of flexible customer information have become a major concern for model offloading.

To reduce computational offloading, this paper proposes the following:

- Edge computing binds the application directly to the user's computer to reduce data browsing.
- 5 G development branching network allows unique tasks to be distributed individually.
- Dynamic algorithm for machine discharge that helps to determine the right path and demonstrate the balance of Nash equilibrium.

MOTIVATION

Different remote systems are interconnected in 4G networks in order to help transfer control from one innovation to the next. In this way, the remote frameworks were planned autonomously and geared toward various types of assistance, information rates, and clients, and consequently, they required a shrewd interworking approach. A significant test for the advancement of 4G is creating compelling, secure, and competent tasks and executives. During a handover, both the portable client and the interconnected remote systems may play a significant role, ensuring administrative congruency and quality. They may also help provide the best assistance to the client. For the 4G plan to succeed, there are several research challenges that should be addressed. The following problems are listed

A System Discovery

4G – multi-mode, multi-get-to, and Reconfigurable network gadgets. This means that each terminal can run more than one type of program at a time and that different applications can be run at different times on different systems. When faced with a situation of this sort, a terminal is given the option of searching for compatible systems. At present, as a solution to this question, Procedure is proposed to be a basic radio-defined program. Parts that were executed in equipment in this strategy are rather actualized using programming on a computer or other processing gadgets that have been installed.

Communication Innovations

4G-organize is a heterogeneous remote situation that requires a range of radio developments, and may have included a radio. Mobile Consumer The need to move between systems to ensure the congruity of administration and improve the standard of assistance. The 4G framework focuses on managing heterogeneous connectivity developments. Increasingly, taking into account the general practicality of the system to meet the Quality-of-Service requirements of the existing assistance.

Device Design

4G is a rapprochement of heterogeneous remote networks. Such systems often rely on different system models and conventions for transport, steering, executive flexibility, etc. Fundamentally interconnecting these devices another examination challenge is to promote involvement between them. System conditions: Network conditions can fluctuate across remote systems, for example, data transfer capability, deferral, jitter, etc., resulting in different qualities of assistance being provided. Flexible consumers maintain standards of administration when crossing heterogeneously while handling a variety of arrangements.

Charging and Billing

As per the 4G arrangement, numerous specialist organizations will be engaged during the meeting if the clients switch from one specialist cooperative system to at least one other specialist cooperative system. Consequently, an individual meeting may include more charges. Other than that, there are likely to be numerous billing plans offered by different administrations. To test this, track the charges per use per portion of a meeting that used the system, administration, or substance. To ensure complete administration, there needs to be more charging arrangements between the specialist cooperative systems.

An Enormous Number of Administrators

There are countless system administrators who work together to coordinate the 4G-internet. Under such circumstances, versatile customers who are responsible for the choice of handover will require higher degrees of control over how administrations should be ensured. This will be combined with versatile trust ties between administrators of organizations.

Security

The degree of protection provided is extraordinary in different systems. Greater interconnectivity and work-to-work would make defenselessness considerably more popular. Wired systems make it extremely difficult to identify, inspect, and prevent worms and infections, but combined wired, local, and transferable systems will pose significantly more difficulty and testing.

Clog Control

Another basic issue of 4G - systems is congestion control. Shirking or avoidance of the blockage and discovery and recuperation after a clog are two fundamental methodologies taken towards blockage control. The evasion strategy would allow the program to conduct control and booking strategies on confirmation fairly. The discovery and recovery will allow executives to trade in stream control and criticism.

The current work on the calculation of the flooding issues has accomplished some positive outcomes. However, they have not consolidated the accompanying indispensable issues. To begin with, they don't consider the information preparation assignments and the mining undertakings together in calculation offloading. Instead, they just examine the calculation offloading issue independently either for information preparation assignments or for mining errands. In any case, practically speaking, these two kinds of errands should be handled at the same time in blockchain-engaged IIoT. In this way, the calculation offloading arrangement must consider the information handling undertakings and the mining errands together to accomplish asset streamlining all around. Second, the greater part of this work accepts that all IIoT gadgets can be associated with the ESs legitimately by means of remote systems. Be that as it may, in certifiable situations, few IIoT gadgets may encounter the lowest availability or may even neglect to interface with any edge passage (AP) purely on the grounds that they are far from the APs. These gadgets at that point need to be associated with very much associated neighboring gadgets to offload assignments to the ES, i.e., a multi-jump calculation offloading situation, in which an errand may be transmitting more than one gadget from the first gadget to the ES. Consequently, certain IIoT gadgets may need to transmit their own assignments and errands from adjacent gadgets, creating the problem of offloading calculations increasingly perplexing. In this paper, we propose an agreeable and distributed multi-bounce calculation offloading solution to find the information handling undertakings and the digging assignments for blockchain-enabled IIoT together.

PROPOSED FRAMEWORK

The framework is comprised of processing, networking, and data management with data from IIOT. The proposed system consists of two parts: establishing a 5G structure and implementing dynamic computational offloading. 5G structure is used because it gives immense support for data offloading. The cloud is linked with a 5G network. The 5G cloud is edge servers will allow control of more devices efficiently, such as control of complex equipment, self-driving cars, enhanced health care, military operations, space exploration and many more uses yet to be found in the future. Data from the user equipment and sensors is collected by the edge nodes in our network and sent to the server. The system's computing design provides a three-layer framework layout that comprises the main 5G network server, local cloud server, and data collection.

The main server is linked via the local server layer. It requires two modules, the normal data interface and the other interface for data maintenance. This layer is where reprogramming or even automatic updating of nodes occurs via dynamic data collection. This layer primarily consists of (i.e.,) application types. Real-time activities run on that layer too. The edge server or local server or second layer links the user equipment and nodes to the local edge server. Almost all networking programs are backed up but the cloud service can be reduced or modified depending on the configuration file.

Edge cloud technology provides the conventional advantages of computing offloading capacity and performance. The computational offloading is performed in the edge cloud using the Dynamic Computation Offloading Algorithm. The algorithm uses the concept of game theory in which a centipede game is used for user equipment (UEs) is considered self-seeking in the offloading sense of multi-hop computing, so that they can make an optimal option move based on their own desires after observing other devices' decisions. As a result of UE's efforts, the company's overall expenses will increase, and the company will find an effective method of transferring data. Therefore, we consider that the IIOT plays an uncooperative game, and Nash equilibrium between the user equipment is achieved by the formula through which each IIOT strives to achieve the maximum possible optimum outcome.

First, each user equipment will initialize its decision and determine the forwarding price for the data transferring on the basis of its rate of cost. Free players are user equipment who don't support other UEs transfer info, the rest are players that are locked.

Second, a time slot synchronization system is implemented for all UEs. All UEs receive the start message through the controller. Once the time slot starts, all UEs receive the start message. During this process, the following functions are performed:

Each unit sends a message with its transfer cost, forwarding cost, and number to the equipment in the next layer after receiving a message, and each free user equipment decides the appropriate actions for the next slot after getting adequate information, but will not update their own decisions directly.

The manager checks the action and sends an authorization request to EUs. When no directing message is obtained during a given time span, the controller transmits an ending message to all user equipment, signaling that the Nash equilibrium is hit.

The status of each UE will be changed at the completion of the slot cycle. 5G calls for network slicing capable of operating several configurations at the same time, while NFV enables regular equipment setup and re-configuration for various purposes rather than direct equipment installation.

All programs and data can be regional for local IoT installations. Apps will only execute efficient functions as necessary, so the data itself can be temporary and not need to be retained for a long time.

Rather than running applications centrally and returning all data to the main server or first layer, the data is stored remotely or in small data centers dispersed through 5G networks. Crowdsourcing helps local network providers to implement macro towers that would be part of the network of the ultimate operator.

When utilizing their towers, these investors must be registered, certified, controlled, and paid immediately. For a seller network operator, sharing assets in 5G is an obvious opportunity, particularly if they offer a subset of these towers with mobile facilities. These two frameworks are known as active sharing.

The MNO incorporates elements of the Radio Access Network as active elements of free sharing. An MNO who shares the building in which the micro tower antenna is located, refrigerators, accommodation, and telecommunications rooms are not considered active sharing.

The 5G network slice is done on the cloud server. Network slicing enables operators to use the same cloud server and serves a broad range of customer devices and apps.

Merits of Proposed System

- The decision engine is the reason for uploading to the cloud.
- The engine retrieves device and application profiler data and applies some logic to it.
- The suggested program completely follows a decision algorithm and is fully automated.
- The engine uses the execution of past activities to estimate their execution period and performance scale.
- Predictions are only feasible because there are historical documents.

System Requirements

Software Requirements

- Python
- Mysql
- Liclipse

Interface Requirements

- User Interface:
- LiClipse

Hardware Interface

- Picocells and Distributed Antenna Systems (DAS)
- MIMO Antenna

Software Interface

- Python
- Flask Framework
- Flask startup and configuration

Communication Interface

- 5G new radio spectrum

System Architecture

In this proposed research the computational offloading is done through a dynamic computational algorithm and 5G structure. 5G's important components of the structure are the 5G core network, 5G spectrum, local server, macro tower, Pico cells, and the end-user. Because of this structure computational offloading is done as, the capacity of local servers is high so latency is reduced because all data can be available in local server, macro tower uses the MIMO technique which allows many input and output at the same time, this technique allows more activity and reduces energy and increases efficiency, pico cells are used as the last structure to connect user equipment, it increases data throughput and also the coordination of pico cells with macrocells through a heterogeneous system can be valuable in

Figure 1. System architecture

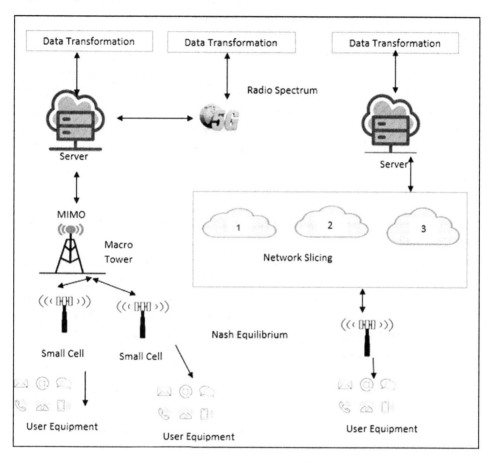

Figure 2. Offloading architecture

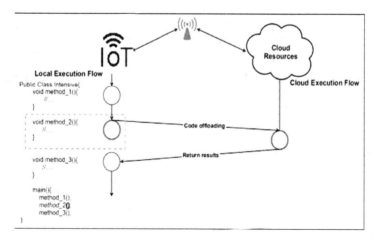

consistent hand-offs and expanded versatile information limit. As the 5G structure provides network slicing due to its higher capacity, edge computing is an advantage. The 5G structure acts as a structure to connect user equipment with the server. Computation offloading can be done at the data management level. This can be done through dynamic computational offloading, which uses game theory and Nash equilibrium. Game theory is used for taking decisions. The centipede game is a type of game theory approach to making decisions. Nash equilibrium is also used for decision making. Dynamic computational algorithm increases energy, data, and time efficiency. Figure 1 shows the architecture of the proposed framework.

Offloading Architecture

Offloading calculation for remote execution is a thought that has been concentrated from that point forward the PC arrangement was created. The calculation offloading innovation points at relocating the calculation escalated code from asset restricted machines to remote-figured assets to quicken the running procedure of the calculation and lessen the vitality utilization of constrained battery gadgets. Figure 2 shows the general idea of offloading. In a general offloading model, calculation serious codes of a portable application are distinguished from the outset, and afterward, it is assessed by the dynamic procedure dependent on the goal of the versatile cloud increase administration (e.g., sparing vitality) to whether offload or not. Lastly, the code is offloaded to the remote processing assets using various sorts of accessible strategies. The code offloading system accepts the entire application calculation that initially occurs on devices. It empowers devices to relocate a portion of the calculation from asset-constrained versatile gadgets to asset-rich registering machines, which makes it adaptable regarding out-sourcing. Be that as it may, the detriment of offloading is that it requires designers to recognize and parcel the piece of the calculation to offload, which is a not-insignificant task and may force pointless overhead for devices. Different offloading approaches have been proposed and applied dependent on different programming models of versatile applications. The methodologies can be characterized into four classes: parcelled offloading, VM relocation, versatile specialists-based offloading, and replication-based offloading.

Algorithm

Pseudocode: Dynamic Computation Offloading Algorithm

- Step 1. Set the TConstraint and the Transmission Rate from the energy and time matrices.
- Step 2. For j=1 to N

- Step 3. Using the random bit stream generator
- Step 4. Specify the starting cell in the table by checking the first bit.
- Step 5. For j=1 to N-1
- Step 6. In the table, place each bit in the proper position for the bi stream.
- Step 7. To determine the total energy and time of each cell, calculate the self-energy and time of each cell.
- Step 8. Compare the new Total Energy of this cell with the previous one in the table if this specific cell has been visited before.
- Step 9. IF the new Total Energy of the cell is less than the previous one.
- Step 10. Substitute the new calculated amount for the energy and time in this cell.
- Step 11. Based on the new amount in this common cell, update the remaining amounts in the remaining cells of the previous bit stream.
- Step 12. To determine the energy and time of the remaining bits of the new bit stream, use the formula below.
- Step 13. Track the position of each bit in the table in a matrix.
- Step 14. Else
- Step 15. Remain in the cell with the previous total of energy and time.
- Step 16. On the basis of the current cell's amount, calculate the energy and time for the remaining cells.
- Step 17. Maintain a matrix that tracks the positions of every bit in the table.
- Step 18. End if
- Step 19. End if
- Step 20. End for
- Step 21. If ((No of bits ==N) & (Etot<Emin&Ttot<Tconstraint & hamming distance criteria))
- Step 22. Return Etot, Ttot
- Step 23. End if
- Step 24. End for

This research proposes an innovative complex programming mathematical formula. Complex coding is an interoperability approach that makes a complex issue a less complex set of problems. Our algorithm is designed for repeatability, i.e., we produce discrete 0s and 1s strings sporadically, and use substrates while enhancing the solution (for example genetic optimization. In addition, we create a new strategic delay on specific substrates in the additional computation with a dynamic programming table. The algorithm can find an almost optimal solution after a few iterations. The conditions for the termination of the hamming distance will be fulfilled by the use of a Hamming distance criterion in exchange to abort the pursuit promptly for the final decision. We found an IoT gadget with N-independent industries that can be run locally or carried on to the cloud. Expect a 5 G-network to be fully accessible for the gadget, but the bandwidth of network transmission can shift dynamically. We also expect the bandwidth will shift in time. The bandwidth of the network will change. Interference and network traffic jams can usually change the bandwidth of the network transmission dynamically.

Based on the current bandwidth of the remote network, the IoT gadget will decide whether to prepare each entity domestically or offload it. Since errors are executed within a time limit, it takes a long time to pass a command from a gadget to the cloud via a connection. When making a decision, the dynamic programming algorithm will take into account the existing bandwidth of the remote network. As we approach the deadline for running the process, we will probably decide which errors should be discharged to the cloud server in order to save energy. The N-by-N table indicates which tasks should be discharged to the Bit streams (where N is the number of entities). A random piece stream is used to determine the primary solution. Assigning 1s to the next lateral cell and 0s to the next linear cell is the purpose of this stream in the table. The starting cell in the primary stream is (1, 2) when the mainstream bit is 1, and it is (2, 1) when the mainstream bit is 0. This strategy prevents the increasing part strings from being affected by additional calculations. As the slice stream is generated at random, we estimate the energy and time consumed by each cell (e.g., each undertaking) in the table and at the same time measure the full energy and execution period of this bit stream. We calculate the energy of a random bit stream up to the primary specific cell in the table and then compute the energy of the current all-out string in this row if the random bit stream was created with any specific cells in the table with an established string. If the current all-out energy of this specific cell is not exactly the former one, we will keep the existing substring and delete the previous substring, supplanting the all-out energy and cell resources of this cell with different amounts. According to the new quality in this particular cell, we adjust the energy and execution time for the remaining cells in the component source. In any event, if the energy of the original component stream is different from that of the current component stream, a specific technique must be played while keeping the current stream. The solution will be terminated and recognized if the Hamming distance is greater than the defined limit from each of the 1 source. The last component of the 1 stream denotes all components that are performed locally.

Nash Equilibrium

Nash balance is a principal idea in the hypothesis of games and the most broadly utilized technique for anticipating the result of key cooperation in sociologies. A game (in vital or ordinary structure) comprises of the following three components:

a lot of players, a lot of activities accessible to every player, and as a result work for every player. The result capacities talk to every player's inclinations over profiles, where an active profile is just a rundown of activities, one for every player. An unadulterated technique Nash balances is an action profile with the property that no single player can acquire a higher result by going amiss singularly from this profile. This idea can best be comprehended by taking a gander at a few models. Look from the start as a game including two players, every one of whom has two accessible activities, which we call A and B. If the players pick various activities, they each get a result of 0. If the two of them pick A, they each get 3, what's more, if the two of them pick B, they each get 2.

This "coordination" game might be spoken to as follows, where player 3 picks a line, player 4 picks a section, and the subsequent adjustments are recorded in brackets, with the primary segment relating to player 3's result: The activity profile (C, C) is a balance since a one-sided deviation to A by any one player would result in a lower result for the straying player. So also, the activity profile (B, B) is additionally balanced. As another model, consider the game "coordinating pennies," which again includes two players, each with two activities. Every player can pick either heads (H) or tails (T); player 3 succeeds a dollar from player 4 if their decisions are the equivalent, and loses a dollar to player 4 if they are most certainly not. This game has no unadulterated procedure Nash equilibria. Now and again, rather than picking an activity, players might have the option to pick likelihood conveyances over the arrangement of activities accessible to them. Such randomizations over the arrangement of activities are alluded to as blended systems. Any profile with blended techniques instigates a likelihood appropriation over activity profiles in the game. Under specific suppositions, a player's inclinations over every single such lottery can be spoken to in a capacity (called a von Neumann-Morgenstern utility capacity) that relegates a genuine number to each activity profile. One lottery ticket is liked to another if and just if it brings about a higher anticipated worth of this utility capacity or anticipated utility. A blended methodology Nash-harmony is then a blended technique profile with the property that no single player can acquire a higher estimate of anticipated utility by veering off singularly from this profile.

Game Theory

There are numerous applications of game theory, such as in business, finance, political theory, and neuroscience. The best way to develop one's thinking and dynamic abilities in an unpredictable world is to know and understand game hypothesis methods, both the well-known ones and a few of the moderately lesser-known ones.

Table 1. NE model 1

	A	B
A	(2,2)	(0,0)
B	(0,0)	(1,1)

Table 2. NE model 2

	H	T
H	(1,-1)	(-1,1)
T	(-1,1)	(1,-1)

Centipede Game

It is a wide structures game in which, conversely, two players have an opportunity to share a slowly growing hoard of cash. Since players make their moves sequentially rather than simultaneously, the centipede is a consecutive game; every participant is aware of the strategies picked by players that came before them. When a player takes the reserve, that player gets the larger piece, and the other player gets the smaller one. Consider the following example: Player A will go first, and he will need to decide whether he wants to "take" or "pass" the reserve, which currently amounts to $2. When he takes, A and B each get $1, but when A passes, it is Player B's decision to take or pass now. If B takes it, she gets $3 (for example, her past reserve of $2 + $1) and A gets $0. In any case, A now has the opportunity to decide whether to take or pass if B passes. Each player gets $100 if they consistently pass the game.

If A and B as shown in Figure 3 and 4, coordinate and keep on going until the end of the game, then they will receive the most extreme payout of $100 each. However, if they doubt the player and anticipate that they should "take" the primary chance, Nash balance predicts that they will take the least conceivable case ($1 in this case). In reality, exploratory investigations have been conducted only from time to time, regardless of this "discerning" behavior as predicted by the game hypothesis. The underlying payout for the last payout was relatively small so this isn't surprising. Similar behavior has been seen by test subjects in the explorer's problem as well.

Advantages of Proposed Algorithm

- Using the previous time execution to estimate the duration and performance size of the execution.

- Using engine in tandem with the network profiler to approximate the runtime of the function in local and cloud contexts.
- Feed data to decision engine: Perform activities on both local and remote servers' concurrent mode for newly developed applications.
- Depending on the output of the decision engine-optimistic mode, run the task only in one context.
- Run when there are no previous documents open at the same time.

STATISTICAL ANALYSIS AND SIMULATION EXPERIMENT

The system is designed in a way that reduces computational network load. It has simulated using Dynamic Computation Offloading Algorithm. Three approaches are used to achieve so. The server is connected directly to the user's equipment using edge computing is the first approach. The second approach allows performing specific tasks independently by utilizing network slicing in 5G emerging technologies. Third, help you find optimum paths and demonstrate Nash balance by the dynamic computational download algorithm. Software upload can be achieved by utilizing these tools. This obviously means data security, latency reduction, and energy consumption reduction too. Computational offload is done using these methods. The system also offers data protection, improved processing speed, latency reduction, and decreased power consumption.5G offers a performance ranking for MIMO. Network slicing also provides data protection by feature separation. Figure 5 shows the computation graph of cost and distance. Hence cost of the proposed system in averaged when in hop distance has increased.

Figure 6 shows the time and transmission rate for the all the possible instance of the network usage. Figure 7 shows the increase in the time when increase in the task. This design can operate up to 50+GHz and can shape up to 10+ layers of the digital beam. The low rate of latency is 1ms. Network energy consumption reduction of such 90%. Enhanced IoT power equipment battery life for up to 10 years.

CONCLUSION AND FUTURE WORK

Edge servers participate in this three-layer cooperative processing system, just as hubs collaborate on edge servers. Our current efforts are directed towards optimizing the offloading options and the distribution of the computation activities in order to reduce the usual scope of activities. The problem mixes geometry optimization with non-geometry optimization as the network size increases and becomes more challenging to solve. We have been studying multi-hop offloading as part of our

Figure 3. Cost versus distance

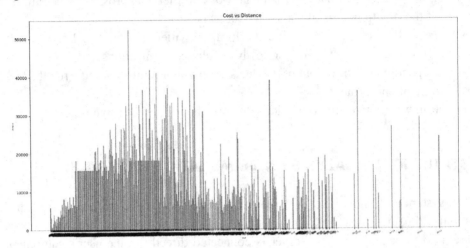

Figure 4. Time versus transmission rate

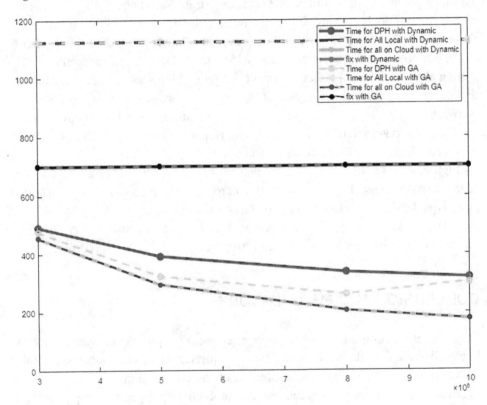

Figure 5. Time versus number of tasks

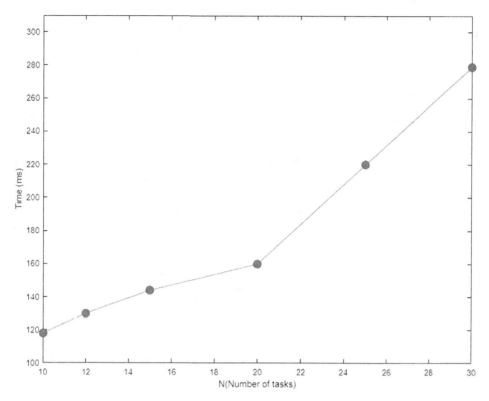

research on edge computing enhanced by 5G. A multi-hop offloading game was formulated as an offloading game and proven to exist in game theory. A high-performance distributed algorithm was developed using this game theory. The development of 5G IIoT facilities will enable the delivery of critical updates to entire networks without overheating, freezing, or overloading computers. The development of 5G and the advancement of IIoT is imminent. A great deal of work needs to be done in this region. In terms of coverage first, 5 G is undoubtedly limited to a particular area similar to its predecessors. Infrastructure is also costed by undergrowth and comes at a cost to the economy. For a small industrial region, it is one of the most expensive to install. There are also concerns regarding privacy: blockchain technologies can tackle privacy problems since standard encryption is not foolproof as a means of securing data.

REFERENCES

Bonomi, F., Milito, R., Zhu, J., & Addepalli, S. (2012). Fog computing and its role in the Internet of Things. *Proc. 1st Ed. MCC Workshop Cloud Comput.*, 1316. 10.1145/2342509.2342513

Chen, M., Hao, Y., Li, Y., Lai, C.-F., & Wu, D. (2015, June). On the computation offloading at ad hoc cloudlet: Architecture and service modes. *IEEE Communications Magazine, 53*(6), 1824.

Craciunescu, M., Chenaru, O., Dobrescu, R., Florea, G., & Mocanu, S. (2020). *IIoT Gateway for Edge Computing Applications.* . doi:10.1007/978-3-030-27477-1_17

Li, Y., & Yang, R. (2021). Edge Computing Offloading Strategy Based on Dynamic Non-cooperative Games in D-IoT. *Wireless Personal Communications*. Advance online publication. doi:10.100711277-021-08891-5

Wang, X., Xia, F., & Rodrigues, J. (2019). Joint Computation Offloading, Power Allocation, and Channel Assignment for 5G-Enabled Traffic Management Systems. *IEEE Transactions on Industrial Informatics.* . doi:10.1109/TII.2019.2892767

Xu, X., Chen, Y., Zhang, X., Liu, Q., Liu, X., & Qi, L. (2019). A blockchain-based computation offloading method for edge computing in 5G networks. *Software, Practice & Experience.*

ADDITIONAL READING

Chen, M. H., Dong, M., & Liang, B. (2018). Resource sharing of a computing access point for multi- user cloud offload- ing with delay constraints. *IEEE Transactions on Computers*, 1–1.

Du, J., Zhao, L., Feng, J., & Chu, X. (2018, April). Computation offloading and resource allocation in mixed fog/cloud computing systems with minmax fairness guarantee. *IEEE Transactions on Communications, 66*(4), 1594–1608.

Lin, L., Liao, X., Jin, H., & Li, P. (2019, August). Computation Of- floading Toward Edge Computing. *Proceedings of the IEEE, 107*(8), 1584–1607. doi:10.1109/JPROC.2019.2922285

Tran, T. X., Hajisami, A., Pandey, P., & Pompili, D. (2017, April). Collaborative edge computing in 5g networks: New paradigms, scenarios, and challenges. *IEEE Communications Magazine, 55*(4), 54–61. doi:10.1109/MCOM.2017.1600863

Chapter 11

Quantum Communication Is the Next Level of Wireless Networking and Security

Sumit Dhariwal
Manipal University Jaipur, India

Avani Sharma
Manipal University Jaipur, India

ABSTRACT

A new technology approach for improving the efficiency of BB84 protocol for Q-Manet is proposed in this chapter. One of the valuable factors in quantum cryptography is that it ensures vital protection. This chapter explains that the quantum communication for Q-Manet can be successfully sent through both quantum channels and classical channels. The tradition of quantum channel technique for quantum communications on Q-Manet are analyzed, which ensures the way to remove the practical difficulties of secure communication. This chapter provides a new mechanism for improving the communication and thereby improving the efficiency of BB84 protocol quantum communication for Q-Manet.

1. INTRODUCTION

1.1 Quantum Communication Networks

A quantum-tum tries to in the field of quantum communication, move qubits from one processor to another over long distances. The quantum internet can be connected to

DOI: 10.4018/978-1-6684-3921-0.ch011

Figure 1. Quantum communication on Q-Manet

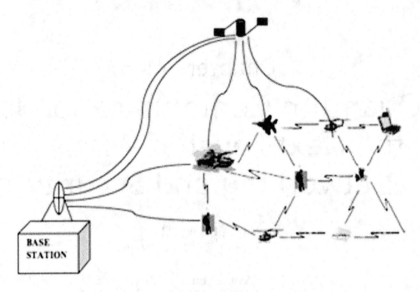

the local quantum network in this fashion. Many applications benefit from a quantum Internet because information may be transported between quantum processors over long distances by using quantum entrained-glade qubits. Most quantum Internet applications simply require relatively basic quantum computers. If these processors can only prepare and measure one qubit at a time, this is adequate In the field of quantum communication, move qubits from one processor to another over long distances. Quantum computing applications, on the other hand, only be accomplished if (mixed) quantum processors are easily accessible and simulated in higher numbers than conventional computers although quantum entanglement between two quantum computers can already be achieved, quantum Internet applications only require modest quantum computers, frequently just a single qubit. In terms of security and speed, traditional computer simulations of entangled quantum systems fall short.

1.2 Network of Quantum Computers

Quantum networks play in quantum computing and communication systems, it plays a crucial role. Quantum networks allow information to be exchanged between quantum computers in the form of physically distinct quantum bits, also known as qubits. A quantum processor is a tiny quantum computer with a set number of qubits that can perform quantum logic gates. Network of Quantum Computers: are comparable to conventional networks in that they operate in a similar manner. Quantum computing, such as quantum networking, is better suited to dealing with specific difficulties like quantum system modeling, as detailed in the following paragraphs.

1.3 Entanglement of Quantum Fields

At this stage, quantum entanglement is a mechanical phenomenon that happens when two or more particles create, interact, or share spatial closeness, even if they are separated by a large distance, causing each photon's quantum state to be influenced by the quantum state of everyone else cannot be characterized on its own. Physical characteristics like distance, velocity, spin, and polarization have been discovered to be linked to entangled particles. For instance, if a pair of particles with a total spin of zero is created, and one of the particles is destroyed, discovered revolving backward on a given axis, the rotation of all other units is also known to be zero. Due to their entanglement, they are measured counterclockwise. This tendency, however, has seemingly paradoxical effects: each measurement of a particle's attribute causes it to irreversibly collapse and alters its initial quantum state. Such a measurement would be made on the entangled system as a whole in the case of entangled particles.

Photons, neutrinos, electrons, buckyballs, and even small diamonds have all been used to illustrate quantum entanglement. The application of entanglement in communication and computation is a burgeoning field of study.

1.5 Processing of Quantum Data

Using Qubits for Computing Quantum information processing is concerned with quantum mechanics-based information processing and computation. Quantum mechanics, unlike current digital machines, are not constrained to different states That what binary numbers are used to encode data (bits). Quantum bits, or qubits, are data encoding units that may exist in many states. Qubits can be constructed of atoms, ions, photons, or electrons, and even the control devices required to operate as computer memory and processors, quantum computers are able to hold multiple of these states at once, they have a natural resemblance. This would allow them to perform some tasks far quicker than any classical computer utilizing the greatest methods available at the time, Integer factorization or quantum many-body simulation are two examples. Quantum computers are still in their early stages of development. The basic building blocks on that path are quantum logic gates and memory, which are based on genuine quantum events like superposition and entanglement.

Figure 2. Photons can be divided into type II photon pairs with mutually perpendicular polarization by the spontaneous parametric down-conversion process.

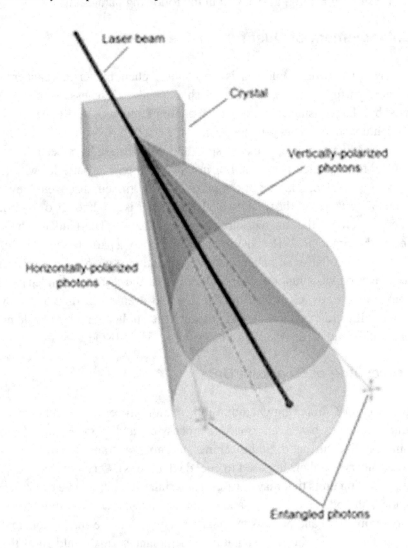

1.6 Teleportation in the Quantum World

Quantum teleportation is a means of sending quantum information (for example, the precise position of an atom or photon) from one location to another via classical communication and transmission. The first shared quantum entanglement between two masses was discovered here. It can't be utilized for faster-than-light transfer or data transfer of classical bits since it relies on classical communication, which

Figure 3. The spooky behavior of quantum bits

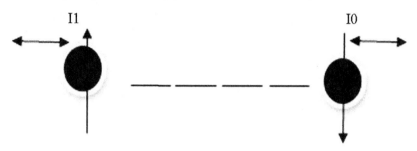

can't go more quickly than the speed of light even though one or more pieces of information have been teleported between two (entangled) quanta, it is not yet complete to be done with anything bigger than molecules. Quantum teleportation is a type of communication that allows a qubit to be transported from one location to another without needing to bring a physical particle with it.

Quantum teleportation requires a teleportable qubit, a normal transmission medium contains a technique for forming an entangled EPR pair and can transfer classical bits (one of four states). Items should each be split into two halves. When one of the EPR pair qubits' quantum state is changed, a and b do Ringing investigations but in the other pair. The steps are as follows:

1. An Entanglement pair is created, with one qubit going to A and the other going to B.
2. EPR pair of qubits bell measurements and transmission qubits is carried out at the location of A, yielding one of four measurement findings that may be stored in two classical bits. Then, at point A, both qubits are discarded.
3. The A typical method is used to send two bits from point A to point B. (After step 1, this is the only potentially time-consuming process due to the speed of light).
4. As a result of the measurement at site A, the EPR pair Qubits at site B are in one of four probable states. The initial quantum state is identical to one of these four states, while the other three are quite similar. The two traditional bits indicate which of the four possibilities was implemented. Knowing this, the EPR pair Qubits at site B were changed in one of three ways, or not at all, to make teleportation qubit-like Qubits.

Figure 4. Qubit-to-qubit quantum teleportation

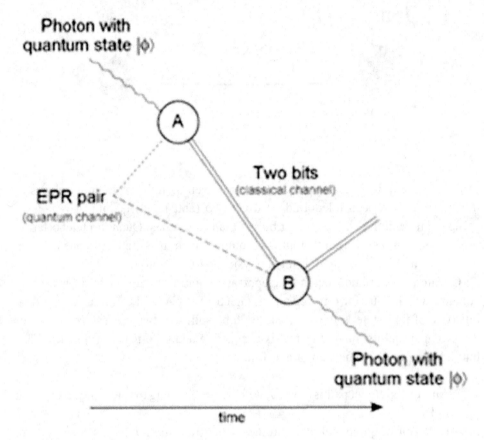

1.7 Protocol BB84

The BB84 protocol was created in the context of photon polarization states and information transfer. Whereas phase-encoded states are used in some optical fiber-based variations classed as BB84, the protocol may be constructed with any second pair of conjugation events.

The transmitter (traditionally known as Alice) a quantum communication channel connects the sender (Alice) and receiver (Bob), allowing quantum states to be transferred, Such a channel is generally an optical wire or an empty area when it comes to photons. They also communicate through conventional public means like broadcast radio and the Internet. The approach is based on the assumption that an eavesdropper (also known as a ewe) can interfere with the quantum channel in any manner they see fit, but the conventional channel must be confirmed. Data is encoded in non-orthogonal states to ensure the protocol's security. The fact that these

Basis	0	1
+	↑	→
×	↗	↘

states can't be measured in general without altering the original state is referred to as quantum. BB84 uses two pairs of states, one conjugated to the other and the other orthogonal to it. The term "bases" refers to orthogonal states in pairs. The vertical (0°) and horizontal (90°) polarization state pairs, as well as the 45° and 135° diagonal bases and the rectilinear bases of the left- and right-handed circular bases, are often used. Because any two of these bases are conjugated, the process can employ any two of them. We'll use rectilinear and diagonal bases in the examples below.

Quantum transmission is the initial stage of BB84. Alice randomly generates bit (0 or 1) and transmits it to one of her two levels (vertical or diagonal in this example). She then constructs a photon polarization-to-state depending on both the bit value and the base, as shown in the attached table. Thus, 0 is represented by a horizontal polarisation state on a rectilinear basis (+), while 1 is represented by a 135° state on a diagonal base (x). Alice sends a single photon to Bob at the specified location using a quantum channel. Alice then records the location, base, and time of each photon transmitted in the random bit phase.

According to quantum physics, no single potential measurement can Since the four polarization states are not all orthogonal, distinguish between them. (particularly quantum uncertainty). Any two orthogonal states are the only comparison that may be conducted (on an orthonormal basis). A horizontal or vertical result can be obtained from a rectilinear measurement, for example. The recti line-AR measurement returns either horizontally or vertically random if the photon was formed horizontally or vertically (as a rectilinear eigenstate); however, if it was created at rectilinear measurements return randomly, either horizontally or vertically, at 45° or 135° (diagonal eigenstate). Furthermore, the photon is polarized to the location ever since this measurement, which was measured (horizontally or vertically), all evidence about its initial polarization is gone.

Bob has no means of knowing what basis the photons were stored on, thus his only option is to measure on a random basis, it can be either rectilinear or diagonal in shape. It keeps track of so that each photon received, the time, measurements base, and measurement result are recorded. Bob and Alice speak over the public traditional connection. once he has measured all of the photons. For each photon, Bob broadcasts the measurement basis, while Alice broadcasts the delivery basis.

Table 1. A quantum communication study in progress connects Alice and Bob.

Alice's bizarre tidbit	0	1	1	0	1	0	0	1
Alice's method of transmitting is based on chance.	+	+	x	+	x	X	x	+
Separation of photons Alice indications a message.	↑	→	↘	↑	↘	↗	↗	→
Bob_s measurement method is based on chance.	+	x	x	x	+	X	+	+
Bob procedures the polarization of photons.	↑	↗	↘	↗	→	↗	→	→
Public Debate on the Premise								
A furtive key that is shared	0		1			0		1
Errors in keystrokes	✓		X			✓		X

Both throw away the photon measurement (bits), but Bob uses a different foundation, half the average. Resulting in a half-bit shared key.

1.8 Quantum Mechanics-Based Cryptography

The application of quantum mechanical phenomena to cryptographic difficulties is the subject of quantum cryptography research. The most well-known application of quantum cryptography is quantum key distribution, which gives a safe solution to the information-theoretically key exchange problem. Quantum cryptography offers the benefit of allowing for the completion of a wide range of cryptographic tasks previously thought to be unachievable with only conventional (non-quantum) communication. Data encoded in a quantum state, for example, cannot be replicated. The quantum state will be altered if an effort is made to read the encoded data (the no-cloning theorem). It can be used for eavesdropping in quantum key distribution. This may be used for listening in randomized access control.

1.9 Dissemination of Quantum Keys

Quantum Key Distribution (QKD) is a dependent communication technology that forms a cryptographic classification using quantum physics. It enables two parties to produce a shared secret key that could be used to encrypt and decrypt communications between them. Because it is the most prominent example of quantum cryptographic uproar, it is known as quantum cryptography.

Quantum Key Distribution is the most well-known and developed use of quantum cryptography (QKD). It's the process of establishing a shared key between two parties (say, Alice and Bob) through quantum communication without such a third party (Eve) learning anything about the key, even if Eve is aware of it. It is possible to listen in on all of Alice and Bob's chats. If Eve attempts to find out what the key is, a dissonance will develop, drawing the attention of Alice and Bob. The key is normally utilized once it has been downloaded. to encrypt conventional communications. Transferred keys can be used for symmetric cryptography, for example. Alice delivers a photon beam to Bob over a quantum channel, such as an optical cable. Each of these photons is polarized in some way, and each one encodes a bit of data (0 or 1). (Either diagonal or rectangular) Bob utilizes a mixture of polarization filters (are these called bases?) and photon detectors to detect these photons. If the polarized photon from Alice goes through his proper filter at his end, Bob will get the matching bit of the photon (0 or 1). A photon has a 50% probability of obtaining a 0 or a 1 if it goes through the wrong filter. After passing all of Alice's photons through his filter, Bob creates a string of 0s and 1s as shown in the below figure.

2. LITERATURE REVIEW

Akihisa TOMITA, Yoshihiro Nambu, and Akio TAJIMA, (2005) provided the latest advances in the Quantum Key Distribution (QKD) technology are discussed. For qubit discrimination at 1,550nm, a high sensitivity photon detector combining the two avalanche photon diodes (APD) has been shown. On a planer light-wave circuit (PLC), a stable interferometer has been created. As a result of the foregoing advancements, single-photon interference has been accomplished across a distance of 150 kilometers. For high-speed (100kbps) key transmission across a 40-kilometer fiber, a temperature-insensitive QKD system is being designed.

Singh, Gutpa, and Singh (2014) presented Quantum Key Distribution (QKD) as a way of distributing secure keys to at least two persons who share them at first. There are several methods for providing a secure key, including the BB84 protocol, the SARG_04, and E91 protocol, and many others this work of all of the related protocols that share a secret key is discussed, and all of the protocols are compared.

Fernandez et al., (2019), In recent years, Many exciting new paradigms for information processing and communication have emerged as a result of quantum information research. Quantum cryptography systems, for example, ensure that information is sent in a totally safe manner thanks to quantum mechanical laws. Such potentially revolutionary security technologies may have important geopolitical and economic implications in the future.

Figure 5. Communication over the public channel

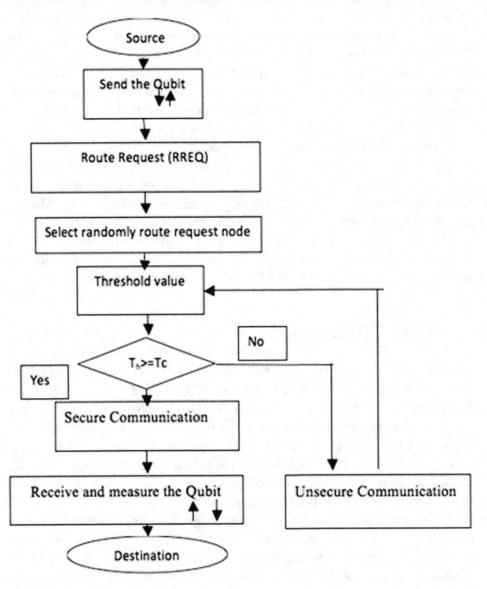

The two primary obstacles for researchers in this sector are transmission range (usually 100 km) and key exchange rate (which can be as low as a few bits per second across lengthy optical fiber lengths). This work adds to the body of knowledge on a way for enhancing critical exchange rates in optical fiber systems with ranges of 1 to 20 kilometers. The results will be provided for a variety of application scenarios, including point-to-point connections and multi-user networks. There have been developed quantum key distribution systems that use standard telecommunications optical fiber and can run at clock rates of up to 2GHz. They present a polarization-encoded variant of the B92 protocol using vertical-cavity surface-emitting lasers with an emission wavelength of 850 nm. Silicon single-photon avalanche diodes with weak coherent light sources are used in single photon detectors. The point-to-point quantum key distribution system has a quantum bit error rate of 1.4 percent and an anticipated net bit rate of over 100,000 bits-1 across a kilometer transmission range.

Rass et al., (2015), Without pre-shared secrets, we offer a mechanism for authenticating the qubit streams being exchanged during the early phases of BB84 quantum key distribution. Unlike the standard technique, which authenticates all protocol messages on the public channel as they are sent, our solution authenticates the qubit stream in advance to validate the peer's identity. We use a second public channel for this purpose, which is physically and logistically distinct from the one used for BB84. It's our alternative to the otherwise unavoidable premise that pre-shared mysteries exist. Moving authentication to an earlier step of the BB84 protocol saves bandwidth during public discussion and makes the system more efficient overall.

Liao et al (2017) Quantum technology lays the groundwork for secure communication by distributing quantum keys (QKD). The fast advancement of QKD over the last two decades has made global quantum communication networks feasible. Small, QKD payloads that can be put on satellites at a low cost of The use of various sizes, such as space stations, is a cost-effective technique to put up this network. This study describes the experimental demonstration of QKD from space to the ground utilizing a small payload from the Tiangong-2 Space Laboratory to the Nanshan Ground Station. The 57.9-kilogram payload includes a tracking device, a QKD transmitter, a synchronization module, and a laser communication transmitter. A 50 MHz vacuum weak decoy-state optical source is broadcast to the space laboratory using a 1 m reflecting telescope.

A 1200 mm aperture telescope at the photon ground station collects signals. To offer a stable and high-transmission communication channel, a high-precision bidirectional tracking system, a polarization correction module, and a synchronization system are employed. When the quantum connection is successfully created, we achieve a key rate of over 100 bps with a communication distance of up to 719 km. When paired with our recent success with QKD in daylight, the present demonstration paves the door for a practical satellite-based global quantum secure network with small-sized QKD payloads.

3. RESEARCH PROBLEMS

Quantum hacking has been defined as a set of assaults that theoretically or technically exploit the security of BB84-based systems in QKD systems. PNS (photon number splitting) assaults fall under the first group. The Intercept Received with Fake States (IRFS) attack, Quantum hacking, which makes use of flaws in avalanche photodiodes (APDs) in electronic detection systems, comes under the second group. We'll go over intercept-receive (IR) assault and intercept-receive (IR) defense in addition to simulated scenarios. (IRFS) attack in the next section.

1.   **Attack (A) Resend Intercept (IR):** Each photon pulse delivered by Alice is measured by the eavesdropper and replaced with another pulse created in the quantum state Alice has so far acquired. in the intercept resist assault. Eve can accurately measure the pulse 50% of the time based on the proper measurement, but Bob picks the same base half as many times; as a result, she produces a 50 percent quantum bit error rate (QBER) = 25% over time (see Figure 1).

Figure 6.

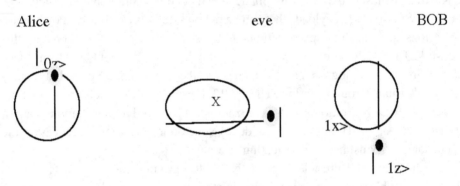

2.   **Attack on Faked States intercept and resend (IRFS):** The eavesdropper false states (IRFS) attack in the intercept received isn't interested in recovering the original states, but rather in creating light pulses that Bob can detect, allowing her to control her as she passes. The quantum channel was undiscovered for a long time. The eavesdropper takes use of faults in Alice and Bob's optical system to trick them into thinking they're When they're truly detecting Eve's light pulses, they're detecting the original quantum states.

Fake states are the name given to these light pulses. Eve can carry out an IRFS assault by exploiting some of Bob's detector's vulnerabilities, such as temporal shift

Figure 7.

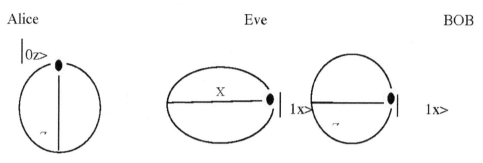

or quantum blindness. An IRFS assault in which the QKD system is controlled by an eavesdropper is known as a quantum blinding attack. Is. To operate APDs in linear mode, strong photon pulses are used. In such attacks, Eve may listen in on the full secret key without boosting the protocol's QBER. Rather than functioning in Geiger mode, the eavesdropper sends light pulses to Bob's station, which are detected by the APD, which functions similarly to a typical photo diode. The IRFS attack may now be used by Eve to obtain the key.

As a response, Bob will get a detection event in the relevant APD detector if he uses the same measuring base as Eve, as shown in the diagram. When Bob uses the opposite base to collect measurements, as shown in Figure 2b, the optical power is uniformly distributed between the two detectors, and no incidence occurs. Eve makes Bob's APD detectors function like standard photodiodes as a consequence. Eve utilizes Bob's public channel announcements to do typical post-processing and get the same secret bits as Alice and Bob in the protocol's final step. A fundamental countermeasure that may be used in electronic detecting systems is a watchdog detector that detects bright simulated states. Simulated states attack and intercept-recent using quantum.

BOB is able to use the same base as Eve:

The basis of BOB and Eve are diametrically opposed and distinct.

To wrap up this section, we want to underline that such IRFS attacker can exploit on a number of QKD protocols, including BB84, SARG04, DPSK, and Consistent One Ways (COW), Eckert, and the decoy state approach. In. Due to Eve and Bob's base mismatch, the attack suffers an extra 3 dB loss. Eve can easily adjust for this in reality by using improved detection capabilities and channel pass losses. Two commercially available QKD systems have attacked detectors without discrimination. Eve is said to have obtained the entire secret key, which is unknown to the parties who are legitimate Finally, when comparing control detector assaults with active base selection to control detector assaults without active base selection, we find

Figure 8.

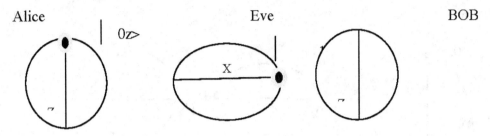

that control detector assaults with active base selection result in half the gain from Eve to Bob.

1. The detector clicks in a predictable manner. when Bob's base choice matches Eve's.
2. In this area for falsified state-run is not discovered when Bob's base choice differs from Eve's.

4. RESEARCH METHODOLOGY

A novel approach for quantum communication via the Q-Manet is provided in this section, which measures the Quan-Tum bit (qubit) process. To forward the qubit to the next node, this technique proposes using the BB84 protocol. Because these are BB84 protocols, the security requirements are quite high as compared to other forms of communication.

The proposed method is based on the approach outlined below, which is broken down into various sections.

*   **Process 1:** The data is transferred to the destination by the source node, which then initializes the sent qubits and transfers them to its neighbor.
*   **Process 2:** Requests for qubits to be routed to nearby nodes are selected.
*   **Process 3:** Route Selection at Random RREQ Qubits should be chosen.
*   **Process 4:** The source node compares the route value to the state of the threshold value to see if it is higher than or equal to.
*   **Process 5:** The communication is secure if the condition is satisfied; else, the information is unprotected.
*   **Process 6:** Receive and measure the Qubits to the destination node.
*   **Process 7:** Finally Qubits safely reached the destination.

Figure 9. Model for quantum communication presented by Q-Manet

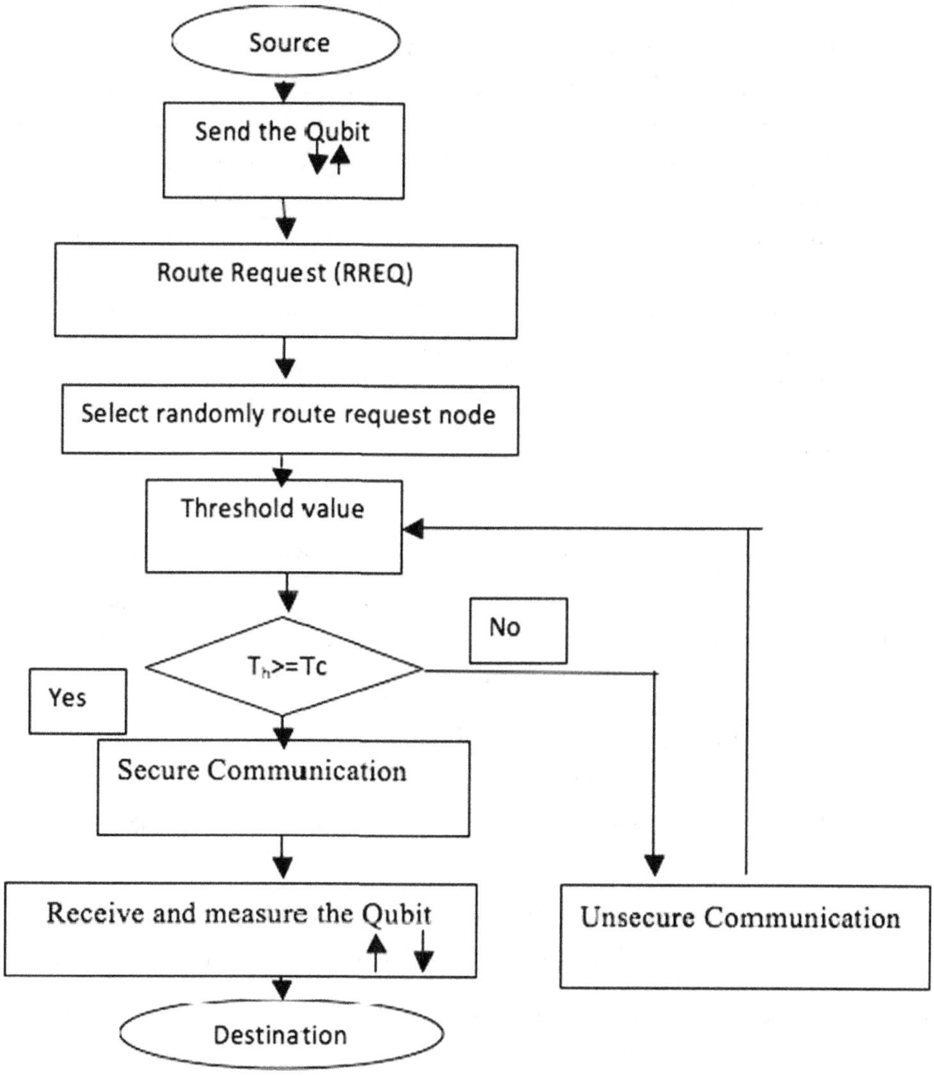

5. STUDY'S SIGNIFICANCE

The goal of this study is to develop a model for the commercial sector, personal area networks, local scale, online banking, tracking applications, intelligent home environments, wildlife monitoring, natural disaster management, military combat zones, air traffic controllers, and real-time communications.

6. CONCLUSION

The use of quantum communication will take us to new heights, allowing us to solve the problem of data transfer delays, as well as the challenges of greater connectivity, flexibility, and cost. Despite their numerous advantages, these technologies are fraught with security risks. As a result, we have highlighted the following points in this paper: Major security challenges that can be exacerbated by quantum communication, we can work in all areas while providing security, and it will greatly benefit us and our network. We will achieve data transfer speeds that we cannot imagine, as well as more services with complete security, thanks to quantum communication. However, with these challenges in mind, the initial deployment design phases will at least make an effort to Remove the possibility of security and privacy lapses.

REFERENCES

Fernandez, V., Gordon, K. J., Collins, R. J., Townsend, P. D., Cova, S. D., Rech, I., & Buller, G. S. (2005, July). Quantum key distribution in a multi-user network at gigahertz clock rates. In Photonic Materials, Devices, and Applications (Vol. 5840, pp. 720-727). International Society for Optics and Photonics.

Liao, S. K., Cai, W. Q., Liu, W. Y., Zhang, L., Li, Y., Ren, J. G., ... Pan, J. W. (2017). Satellite-to-ground quantum key distribution. *Nature*, *549*(7670), 43–47.

Rass, S., Konig, S., & Schauer, S. (2015). BB84 quantum key distribution with intrinsic authentication. *ICQNM*, 41-44.

Singh, H., Gupta, D. L., & Singh, A. K. (2014). Quantum key distribution protocols: A review. *Journal of Computational Engineering*, *16*(2), 1–9. doi:10.1155/2014/785294

Tomita, A., Nambu, Y., & Tajima, A. (2005). Recent progress in quantum key transmission. *NEC J. of Adv. Tech*, *2*(1), 84-91.

Chapter 12

Scheduling Optimization Based on Energy Prediction Using ARIMA Model in WSN

Pooja Chaturvedi
ⓘ https://orcid.org/0000-0001-5207-2696
Institute of Technology, Nirma University, Ahmedabad, India

Ajai Kumar Daniel
Madan Mohan Malaviya University of Technology, Gorakhpur, India

ABSTRACT

Wireless sensor networks (WSNs) have attracted great attention because of their applicability in a variety of applications in day-to-day life such as structural monitoring, healthcare, surveillance, etc. Energy conservation is a challenging issue in the context of WSN as these networks are usually deployed in hazardous and remote applications where human intervention is not possible; hence, recharging or replacing the battery of sensor nodes is not feasible often. Apart from energy conservation, target coverage is also a major challenge. Scheduling the nodes to exist in active and sleep modes is an efficient mechanism to address the energy efficiency and coverage problem. The chapter proposes an ARIMA model-based energy consumption prediction approach such that the set cover scheduling may be optimized. The chapter compares the efficiency of several ARIMA-based models, and the results show that the ARIMA (0,1,2) model provides best results for the considered scenario in terms of energy consumption.

DOI: 10.4018/978-1-6684-3921-0.ch012

INTRODUCTION

WSNs are autonomous network is a collection of sensor nodes deployed in close proximity to perform the environmental phenomenon monitoring through mutual cooperation. These sensor nodes are scattered in remote and hazardous environment without intervention from human is usually not possible for larger duration so the sensor nodes face resource constraints such as limited, energy, bandwidth and communication capabilities. The node scheduling approaches aim to address the network lifetime and coverage which keeps the redundant nodes in low energy state to preserve the energy such that life of sensor network is maximized in I. F. Akyildiz (2002), J. Yick(2008).

The paper proposes a data prediction based approach to optimize the node scheduling strategy by extending the Energy Efficient Coverage Protocol (EECP). In EECP the nodes are divided into subsets to provide surveillance to the specified points with required confidence. The nodes satisfying the required quality of service are in active state and remaining in sleep state. The probability of observation of the node is determined using target detection probability and contribution and trust factor of the nodes. In this approach the network operational time depends on the count of subsets obtained. The proposed approach is a combination of node scheduling and prediction mechanism to optimize the target coverage. The ARIMA based prediction models predict the energy requirement of the nodes in the next round based on the energy requirement of the previous rounds. Different ARIMA models are simulated and analyzed for the energy optimization.

The structure of the paper contains following sections: literature review in section 2, protocol for node scheduling in section 3, and preliminary concept related to ARIMA in section 4, proposed prediction model in next section, section 6 is about results and discussion and conclusion and section 7 contains conclusion.

RELATED WORK

Conservation of energy is a significant researched problem. Most energy conservation approaches are broadly concerned with energy efficient routing, optimal node placement, node scheduling and prediction based strategies. The energy efficient routing approaches enhance the longevity of sensor network by selecting optimal route for the data dissemination. The clustering techniques are also used in conjunction with the routing approaches to reduce the count of data transmissions by designating the data aggregation task to the Cluster Head (CH) nodes. The node deployment strategies are often defined as a multi objective problem to lower the cost and energy requirement. The node scheduling approaches aim to activate only a subset of nodes, hence prolonging the network lifetime. The scheduling approaches aim to identify the coverage redundant nodes and put it into least energy consuming state. If the event's distance is less than the sensing range then it is said to be covered. The data prediction based strategies lower the count of data transmissions needed and hence lowers the energy consumption. The different prediction schemes are used to predict the node state for the next time instant. The prediction is considered valid if the error in prediction lies within the specified error threshold.

Coverage is also a major research problem in context to WSN and is a quality of service metric. Coverage ensures the monitoring interval and quality of the observation of region of interest. Node scheduling based strategy addresses the energy conservation and coverage problem by putting the redundancy of nodes in energy saving state. Various version of the coverage problem are proposed as coverage from energy efficiency perspective, coverage aware deployment, coverage from the connectivity perspective and combined perspective. The coverage problem ensures the possibility of observation of event of interest for longest possible duration and with highest quality whereas connectivity solution describes the path from the sensing nodes to the central node. There have been various approaches proposed to deal with the coverage and connectivity issues. Borasia (2011) and Al-Karaki (2017) proposed a node deployment strategy for addressing the coverage and connectivity issues. Mohamdai(2014) proposed a solution based on learning automata, which determine the next action of the node. Jiang(2012) have proposed a probabilistic prediction based scheduling protocol for energy efficient target coverage approach. Ye. D. (2017) have proposed an adaptive node scheduling.

The most efficient approach addressing both these problem is to activate the nodes according to their energy state and the region under consideration is monitored for the longest possible time and increasing the network lifetime. There are several scheduling based strategies proposed for the coverage problems. The scheduling approaches for area coverage are considered with the monitoring of all the points in the area, but the point coverage approach aims to provide coverage to defined collections of points and the scheduling based approaches for barrier coverage ensure the coverage of the intruder across the perimeter of the target region. The set cover obtained through scheduling based approaches can be categorized in either non disjoint or disjoint. The disjoint set cover based approach invokes the constraint that a node can belong to a set cover whereas the non-disjoint set cover based approaches allows the nodes to exist in multiple set covers. However the inclusion of a node into different set cover is controlled by the initial battery level of the devices (Mullidan (2010), Thai (2008), N. Jaggi(2006), Xiaochun(2006), Z. Zhou(2004), Wang J. (2006), M. Cardei (2005).

The closest related work is in Chaturvedi P. (2015) as EECP, in which a scheduling the nodes based strategy for point coverage using the coverage contribution of the nodes, nodes' trust parameters and probability of coverage detection. The target coverage problem is formulated as ILP. The operation of EECP is divided into three parts as initial, sense the environment and data transmission. The setup phase determines the subset of nodes to cover all the targets. The probability of observation towards a target is derived as a combination of coverage detection probability and trust levels. The central station activates the set covers to appear in the sleep or active state. During sensing phase, the active sensor nodes collect the data about environment through sensing. The sensed data is transmitted to the base station through the hybrid routing strategy. In this approach the objective is to lower the count of nodes in a set cover and enhance the activation time of set cover while ensuring the desired coverage. The network lifetime in this approach is dependent on count of set covers. The high the set covers the larger the network functional time. The motivation of this approach is to optimize the energy requirement in the scheduling strategy using the data prediction techniques as an extension to EECP protocol. The comparison of the proposed mechanism with EECP is in the Fig. 1.

NODE SCHEDULING PROTOCOL

A network consist of u nodes and t targets then the goal is to reduce the energy requirement across each set cover such that the network functional time is optimized in the EECP protocol. Let the count of set covers obtained using the EECP protocol is sc and energy consumption in a set cover i is ec_i then the point coverage is formulated as an ILP problem having the goal to reduce the function:

$$E = \sum_{i=1}^{sc} ec_i \tag{1}$$

such that $p_{jk}^{obs} \geq RCL \forall j \in u, k \in t$ which ensures that required coverage level is achievable for all the points RCL and a_{jk} is a Boolean term which is the activation state of a node.

$$a_{jk} = \begin{cases} 1, & \text{if node } j \text{ is active for target } k \\ 0 & \text{otherwise} \end{cases} \tag{2}$$

The proposed protocol predict the energy requirement of the nodes in the next time instant on the basis of historical time series of energy prediction data using the ARIMA models. If the error in the predicted values is within the specified threshold then the predicted values is considered. Using the prediction based strategy the energy requirement is reduced as the count of communication messages is reduced.

Figure 1. Comparative flow charts of EECP and Proposed ARIMA protocol

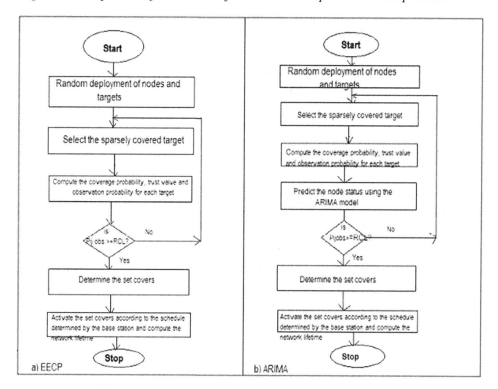

ARIMA MODELS

Due to inherent resource constraints in the WSNs, the prediction models based on time series data is becoming popular. These prediction models are simpler to implement and aims to reduce the energy requirement of the network by reducing the count of transmissions. ARIMA models are prediction model used in univariate time series data for short term predictions. It utilize past observations to predict the short term future values. The accuracy of the ARIMA based models depends on the data sampled.

The various linear approaches used for prediction in WSN are ARMA, AR and ARIMA and MA. The ARIMA based prediction model consists of three components as: Autoregressive components, Moving Average component and an element consist of first derivative of time series.

AR (a) represents the Auto Regressive component of order *a* and is determined as the lag of the time series data:

$$D(s) = \sum_{s=1}^{a} \mu_s D(c-s)$$ (3)

where μ_s represents the weight factors of preceding observations.

MA (b) represents the moving average component of order *b* and is determined as the lag and error in the preceding observations as:

$$D(s) = \sum_{s=1}^{b} \lambda_s \varepsilon (c-s)$$ (4)

where λ_s is the weight factor of different forecasting errors.

ARIMA model combines the characteristics of Auto regressive and moving average components. ARIMA Model of order *(a, b, c)* represents the count of auto regressive terms as a, count of non-seasonal difference needed for stationary data and c is the count of lagged predicted error. ARIMA of order (a, b, c) is determined as the combination of *AR(a)* and *MA(b)* and defined as:

$$D(s) = \sum_{s=1}^{a} \mu_s D(c-s) - \sum_{s=1}^{b} \lambda_s \varepsilon (c-s)$$ (5)

The ARIMA based prediction model consists of three steps (G Box(2008), S Yang(2007))

1. *Preparing the data-* The data is analyzed by decomposing it into different components. The data is further processed to remove the trend and seasonality. The differencing technique is required to make the data stationary.
2. *Diagnosing and estimating model parameters-*The AR and MA components are identified. The best fitted ARIMA is selected on the basis of error, *df* and *p*-values.
3. *Forecast and measure the efficiency of the model-* The optimal parameter values for the next time instant is predicted. The forecast accuracy is determined on the basis of *MAE, MSE, RMSE* and *MAPE* parameters. The performance of ARIMA model is evaluated using following metrics:

Error (E)

Error is the difference between the actual and predicated data values. If the actual data value is a and predicted data value is f, then the error (E) is determined as:

$$E = a - f \tag{6}$$

Mean Error

It is determined as the average of the error of the n predicted values.

$$M = \sum_{i=1}^{n} \left(\frac{a_i - f_i}{n} \right) \tag{7}$$

Mean Absolute Error (MAE)

MAE is defined as average of the absolute error of the n predicted values.

$$MAE = \sum_{i=1}^{n} \left(\frac{|a_i - f_i|}{n} \right) \tag{8}$$

Mean Square Error (MSE)

It is the average of the square value of the error of the n predicted values.

$$MSE = \sum_{i=1}^{n} \left(\frac{(a_i - f_i)^2}{n} \right) \tag{9}$$

Root Mean Square Error (RMSE)

It is the average of the square root value of the mean square error of the n predicted values.

$$RMSE = \sum_{i=1}^{n} \left(\frac{\left((a_i - f_i)^2 \right)^{1/2}}{n} \right) \qquad (10)$$

Mean Absolute Percentage Error (MAPE)

It is the average percentage error of the n predicted values.

$$MAPE = \sum_{i=1}^{n} \left(\frac{\left(\frac{a_i - f_i}{a_i} \right) a_i - f_i}{n} \right) \qquad (11)$$

Q* Error

This measure is used to represent the standard error between the predicted and actual data models.

Degree of Freedom(DF)

The degree of freedom represents a numeric or character variable indicating the approximation of Chi Square statistical distribution.

p-Value

It is used to represent the possibility of correlation with the lags.

The accuracy of ARIMA model is evaluated using the Box-Jenkin Method on the basis of following parameters:

Akaike Information Critera (AIC)

AIC is the information criteria which determines the balance between the actual and predicted data values. It represents the goodness of the predicted model with respect to the actual data values. AIC value is determined as:

Figure 2. Framework of the ARIMA based prediction model

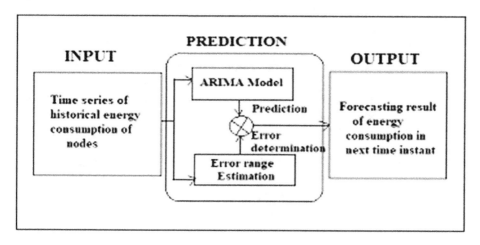

$$AIC = -\frac{2lh}{n} + \frac{2r}{n} \qquad (12)$$

Where, n is the count of observation, lh is the log likelihood and r is the count of right repressors.

Bayesian Information Criteria (BIC)

It represents the posterior likelihood of the match between the predicted and actual data values. *BIC* value is determined as:

$$BIC = -\frac{2lh}{n} + \frac{(r\log n)}{n} \qquad (13)$$

PROPOSED PREDICTION MODEL

In the proposed approach we aim to optimize the energy consumption of the nodes. In this we utilize the past observations to predict the nodes' energy requirement for the next time instant using the Auto ARIMA model. The framework of the proposed model is in the Fig.2.

The proposed model takes as input the energy consumption of previous observations and using the ARIMA based model, the energy need of the next time

Figure 3. Energy graph for set covers

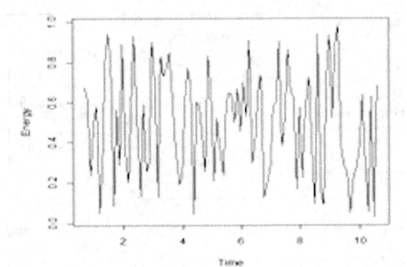

instant is derived. If the error of the predicted value is less to the threshold then the predicted value is utilized in scheduling the set covers.

SIMULATION RESULTS AND DISCUSSION

Data Preprocessing

The first step in modeling the ARIMA based prediction system is to process the data so as to remove the trend and seasonality. The different steps involved in the data preprocessing stage are as follows:

Plot the Time Series Data

The time series representation of the energy need in different set covers is as shown in the Fig.3.

The Auto Correlation Factor (ACF) and Partical Auto Correlation Factor (PACF) plots is in Fig.4.

Decomposition of Data

The different constituents of the energy consumption parameter is shown in the Fig.5.

Figure 4. ACF and PACF of energy consumption

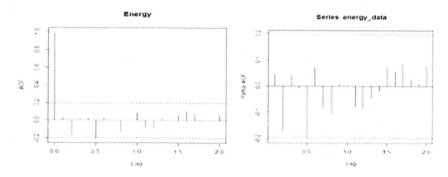

Figure 5. Decomposition of time series data of energy consumption

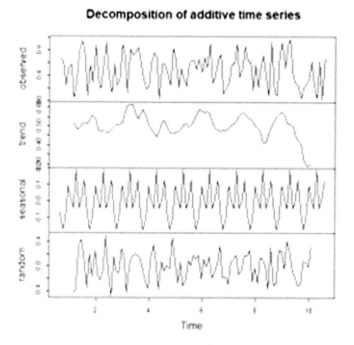

Removing Trend

The time series data is difference to make it stationary. The results from the first and second difference is in the Figure 6.

Figure 6. Differencing time series data to remove stationarity

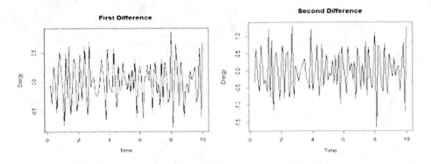

Log Transform Data

The log transform of the data is obtained in the Figure 7.

Figure 7. Log transform of the data

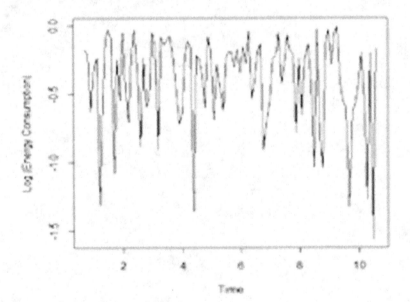

Difference Log Transform Data

The difference of the log transformed data is calculated to remove the stationary on both the mean and variance. The results are in the Figure 8.

Figure 8. Differenced log transform data

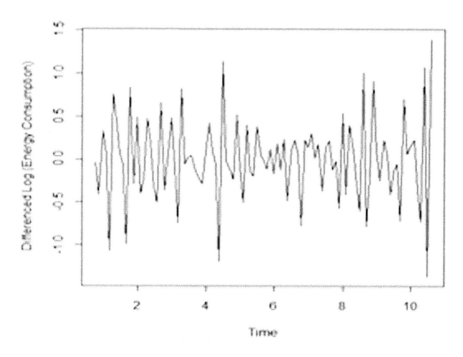

Finding the Order

The ACF and PACF plots are plotted of the obtained data series and identify the order. The results are in the Figure 9. The ACF plot has no frequent spike after lag 0, so order of autoregressive component is 0. The results show that the PACF plot has continuously diminishing plot after lag 2, so the order of MA component is 2.

Estimating Model Parameters

We have compared the performance of different ARIMA models in the Table 1.

Accuracy Measures

The accuracy measures of various ARIMA models are in the Table 2.
It is clear from the above table that the ARIMA (0,1, 2) provides the least error value.

Figure 9. Plot of ACF and PACF

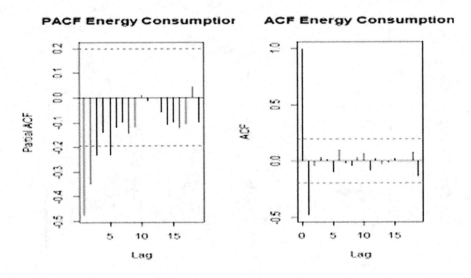

Table 1. Performance metrics for different ARIMA Models

S. No.	Model	AR1	AR2	MA1	MA2	Intercept	Sigma^2	Log likelihood	AIC
1	C(0,0,1)	-	-	0.0440	-	0.5169	0.06662	-5.81	17.63
2	C(0,0,2)	-	-	-0.0773	-0.3093	0.5179	0.06139	-2.24	12.48
3	C(0,1,1)	-	-	-1.0000	-	-	0.06743	-8.53	21.07
4	C(0,1,2)	-	-	-0.9337	-0.0663	-	0.06738	-8.44	22.88
5	C(1,1,1)	0.0317	-	-1.0000	-	-	0.06741	-8.49	22.98
6	C(2,1,1)	0.0305	-0.2658	-1.0000	-	-	0.06227	-5.27	18.54
7	C(2,1,2)	0.1023	-0.2678	-1.0778	0.0779	-	0.06222	-5.26	20.52
8	C(1,0,1)	-0.4205	-	0.5295	-	0.5168	0.06572	-5.21	18.41
9	C(1,0,2)	0.4797	-	-0.6439	-0.3561	0.5181	0.05185	3.8	2.39
10	C(2,0,1)	0.8136	-0.2255	-1.0000	-	0.5181	0.05226	3.4	3.21
11	C(2,0,2)	0.4995	-0.0171	-0.6619	-0.3381	0.5181	0.05184	3.81	4.39
12	C(0,1,0)	-	-	-	-	-	0.1311	-35.86	73.73
13	C(1,1,0)	-0.3483	-	-	-	-	0.1148	-30.04	64.07
14	C(1,1,2)	-0.4054	-	-0.4786	-0.5214	-	0.06646	-7.82	23.64

Based on the AIC value, ARIMA (0, 1,2) is best.

Table 2. Accuracy measures of different ARIMA Models

S. No.	Model	MAE	MSE	RMSE	MAPE
1	C(0,0,1)	0.2711965	0.09333253	0.3055037	0.030068
2	C(0,0,2)	0.2851608	0.10427987	0.3229239	4.190075
3	C(0,1,1)	0.2723514	0.09440541	0.3072546	4.054676
4	C(0,1,2)	0.2706459	0.09283019	0.3046805	4.018337
5	C(1,1,1)	0.2714601	0.09358513	0.3059169	4.036158
6	C(2,1,1)	0.2824570	0.10378674	0.3221595	4.198200
7	C(2,1,2)	0.2840528	0.10491362	0.3239037	4.211797
8	C(1,0,1)	0.2707772	0.09263003	0.3043518	4.006501
9	C(1,0,2)	0.2697746	0.09124237	0.3020635	3.954019
10	C(2,0,1)	0.2686903	0.09007201	0.3001200	3.923560
11	C(2,0,2)	0.2697470	0.09117572	0.3019532	3.951505
12	C(0,1,0)	0.2030828	0.06355327	0.2520977	1.858489
13	C(1,1,0)	0.2016007	0.06248469	0.2499694	1.918775
14	C(1,1,2)	0.2706393	0.09250682	0.3041493	4.003358

Forecasting

The forecasted value of energy consumption of the next 5 time instants are as shown in the Figure 10 and Figure 11 for ARIMA(1,1,2) and ARIMA(0,1,2) respectively.

Diagnosing Residual Series to Find the Best Fitted ARIMA Model

The last step to determine the efficiency of ARIMA based forecasting model is its residual analysis. The residual plots of ARIMA models is in the Table 3.

The results of Auto ARIMA= c (0,0,0) is in Table 4.

The validity of the ARIMA model can be evaluated on the basis of the residual plots. The residuals represent the amount of information left after fitting the prediction model. The good prediction model yields the residuals having following properties:

1. There should be no correlation between the residual values.
2. The mean of the residual values should be close to zero, otherwise the predicted values are said to be biased.

Figure 10. Forecasting using ARIMA (1,1,2)

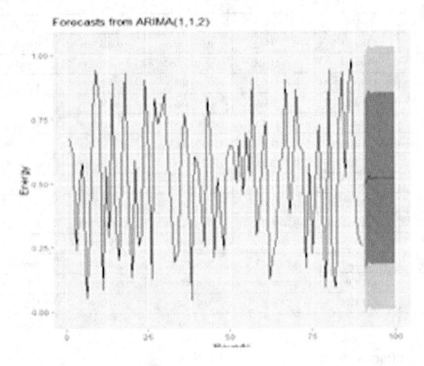

The residual plots of different ARIMA models considered are in the Figure 12 to 25. It can be inferred from the plots that the residual plot for each ARIMA satisfy the properties of the good prediction model.

The plot of ACF and PACF of Auto ARIMA model is in the Figure 26.

The proposed approach's efficiency is analyzed through energy consumption and energy savings for sensor nodes [10]–[50] for constant targets. The standard energy consumption model is used (Heinzelman (2000)).The energy consumption is compared with the EECP in the Table 5 which is lowered significantly.

The comparison of the energy consumption for both the approaches is in Figure 27.

The proposed reduces the number of communications required; therefore the need is lowered. The energy saved is in Table 6.

The results in Figure 28 show that the energy saving is increasing with the proposed approach.

Figure 11. Forecasting using ARIMA (0, 1, 2)

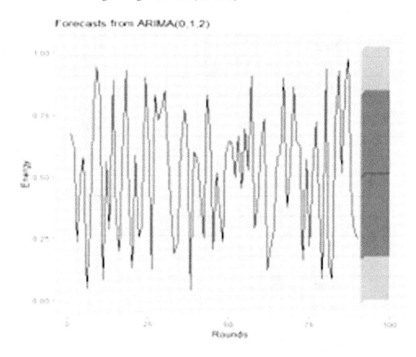

Table 3. Residual plots of considered ARIMA Models

S. No.	Model	Q*	df	p-value	Model df	Lags Used
1	C(0,0,1)	17.286	8	0.02726	2	10
2	C(0,0,2)	11.864	7	0.1051	3	10
3	C(0,1,1)	16.792	9	0.05207	1	10
4	C(0,1,2)	17.32	8	0.02694	2	10
5	C(1,1,1)	17.079	8	0.0293	2	10
6	C(2,1,1)	10.48	7	0.163	3	10
7	C(2,1,2)	10.79	6	0.09508	4	10
8	C(1,0,1)	15.277	7	0.0326	3	10
9	C(1,0,2)	11.53	6	0.0733	4	10
10	C(2,0,1)	14.165	6	0.02784	4	10
11	C(2,0,2)	11.675	5	0.03953	5	10
12	C(0,1,0)	40.736	10	1.257e-05	0	10
13	C(1,1,0)	40.119	9	7.23e-06	1	10
14	C(1,1,2)	15.155	7	0.03406	3	10

Table 4. Parameter values of Auto ARIMA model

Coefficients	Values
Mean	-0.3996
S.E.	0.0340
Sigma^2 Estimated	0.1168
Loglikelihood	-34.03
AIC	72.07
AICc	72.19
BIC	77.28
ME	3.828724e-17
RMSE	0.34007
MAE	0.2689511
MPE	-185.4211
MAPE	215.9662
MASE	0.8383746
ACF1	-0.02863827

Figure 12. Residual curves in ARIMA (0, 0, 1)

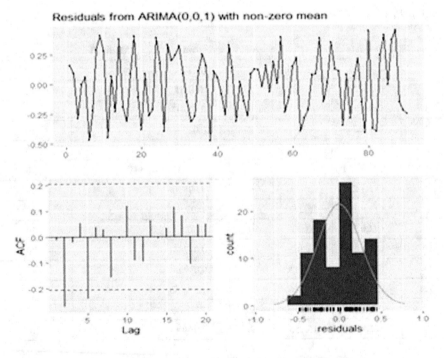

Figure 13. Residual curves in ARIMA (0, 0, 2)

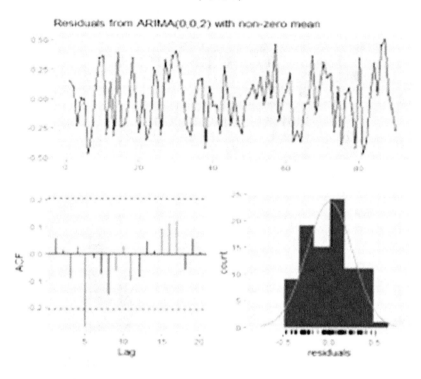

CONCLUSION AND FUTURE SCOPE

The paper proposes an ARIMA based prediction model to optimize the energy consumption of different set covers as an extension to EECP protocol. In this work we have developed and analyzed several ARIMA models using MAE, MSE, RMSE and MAPE. The results show that the ARIMA (0, 1, 2) provides the optimum results in terms of AIC and p values. It is found from the results the proposed approach lowers the energy requirement on increasing the number of nodes from 10 to 50. Thus the proposed prediction technique achieves better energy efficiency. To evaluate the efficiency of the proposed protocol using network functional time and throughput is our future work. The protocol can be applied on a heterogeneous network consisting of nodes having different processing, communication and storage capabilities.

Figure 14. Residual curves in ARIMA (0, 1, 1)

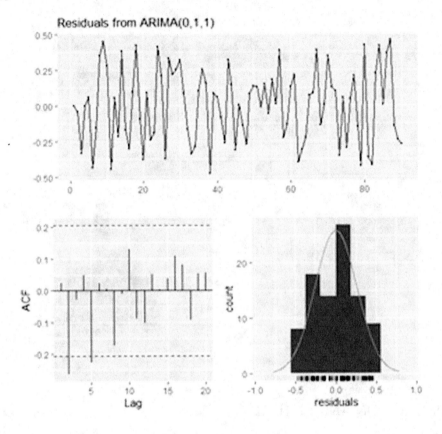

Figure 15. Residual curves in ARIMA (0, 1, 2)

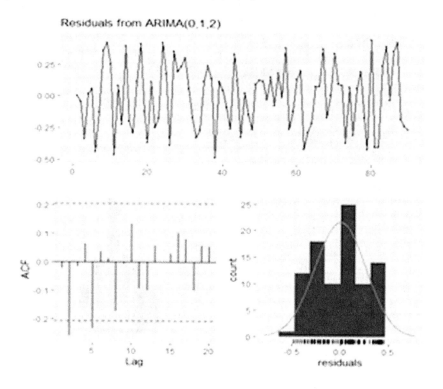

Figure 16. Residual curves in ARIMA (1,1,1)

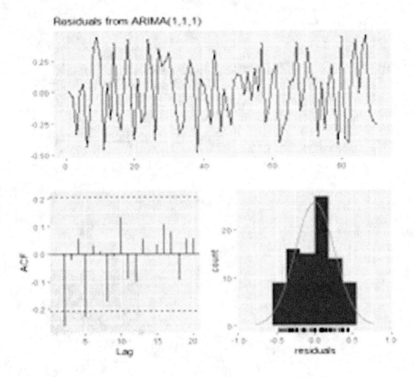

Figure 17. Residual curves in ARIMA (2, 1, 1)

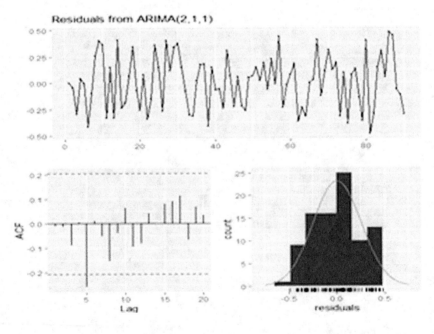

Figure 18. Residual curves in ARIMA (2, 1, 2)

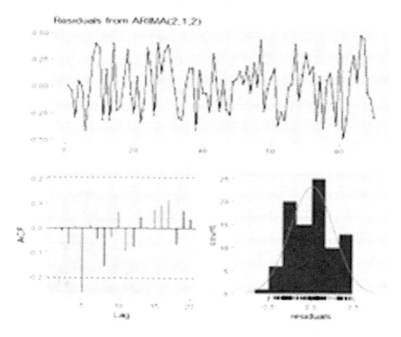

Figure 19. Residual curves in ARIMA (1,0,1)

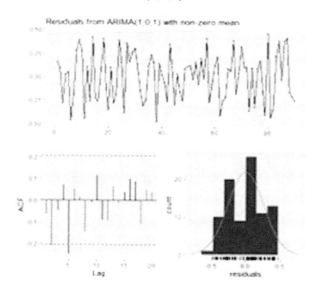

Figure 20. Residual curves in ARIMA (2,0,2)

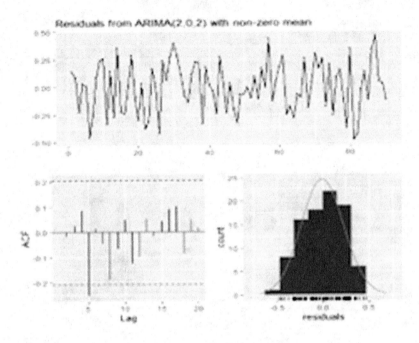

Figure 21. Residual curves in ARIMA (0, 1, 0)

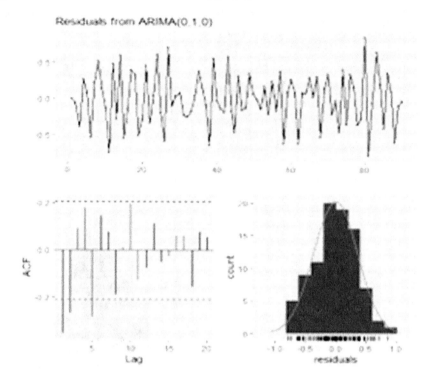

Figure 22. Residual curves in ARIMA (1,0,2)

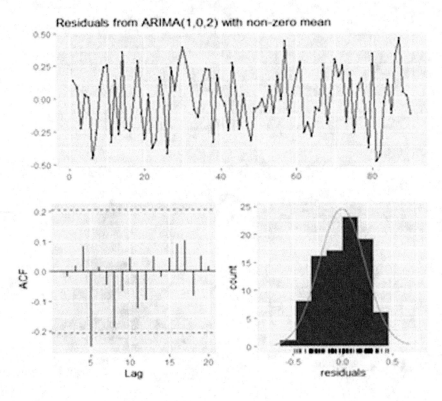

Figure 23. Residual curves in ARIMA (2,0, 1)

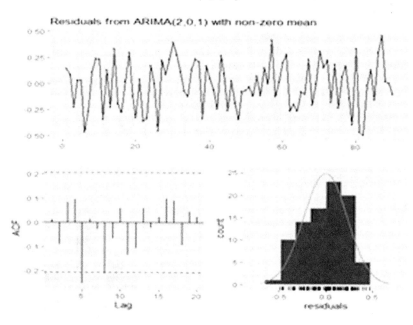

Figure 24. Residual curves in ARIMA (1, 1, 0)

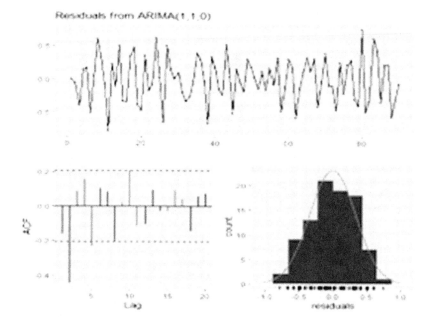

Figure 25. Residual curves in ARIMA (1, 1, 2)

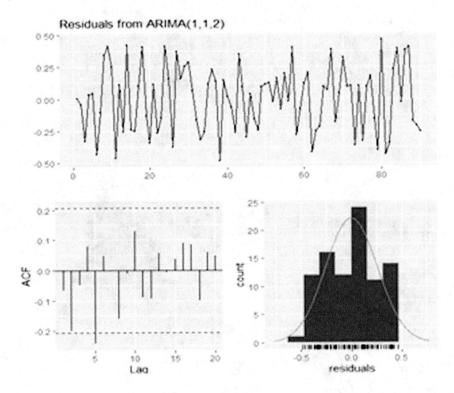

Figure 26. Residuals of ARIMA(0,0,1)

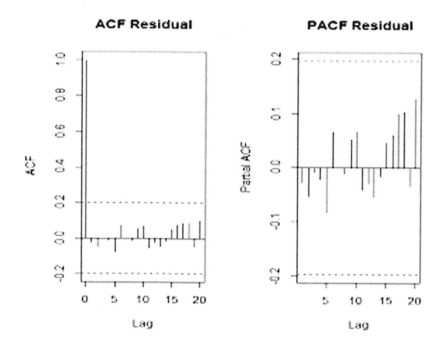

Table 5. Comparison of energy consumption

Nodes	EECP	ARIMA-EECP
10	14	11.2
15	21	16.8
20	28	22.4
25	35	28
30	42	33.6
35	49	39.2
40	56	44.8
45	63	50.4
50	70	56

Figure 27. Energy consumption

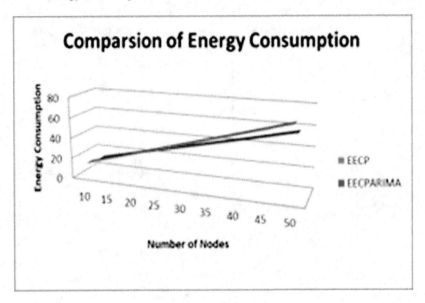

Table 6. Comparison of energy saved

Nodes	EECP	ARIMA-EECP
10	36	38.8
15	54	58.2
20	72	77.6
25	90	97
30	108	116.4
35	126	135.8
40	144	155.2
45	162	174.6
50	180	194

Figure 28. Comparison of energy saved

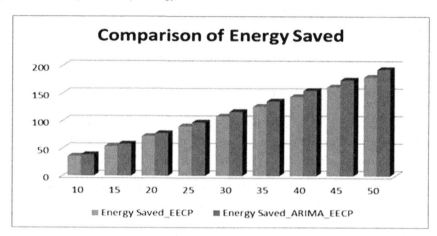

REFERENCES

Akyildiz, I. F., Su, W., Sankarasubramaniam, Y., & Cayirci, E. (2002). Wireless Sensor Networks: A Survey. *Computer Networks*, *38*(4), 393–422. doi:10.1016/S1389-1286(01)00302-4

Al-Karaki, J. N., & Gawanmeh, A. (2017). The optimal deployment, coverage, and connectivity problems in wireless sensor networks: Revisited. *IEEE Access: Practical Innovations, Open Solutions*, *5*, 8051–18065. doi:10.1109/ACCESS.2017.2740382

Borasia, S., & Raisinghani, V. (2011). A review of congestion control mechanisms for wireless sensor networks. In *Technology systems and management* (pp. 201–206). Springer. doi:10.1007/978-3-642-20209-4_28

Box, G., Jenkins, G., & Reinsel, G. (2008). *Time Series Analysis: Forecasting and Control*. Wiley.

Cardei, Thai, & Li. (2005). Energy-efficient target coverage in wireless sensor network. *Proc. Infocom'05*.

Chaturvedi, P., & Daniel, A. K. (2015). An energy efficient node scheduling protocol for target coverage in wireless sensor networks. *5th International Conference on Communication System and Network Technologies(CSNT-2015)*.

Heinzelman, W. R., Chandrakasan, A., & Balakrishnan, H. (2000). Energy-efficient communication protocol for wireless micro sensor networks. In *HICSS '00: Proceedings of the 33rd Hawaii International Conference on System Sciences (vol. 9)*. IEEE Computer Society.

Jaggi, N., & Abouzeid, A. A. (2006). Energy-Efficient Connected Coverage in Wireless Sensor Networks. *Proc. 4th Asian International Mobile Computing Conference (AMOC)*, 77-86.

Jiang, B., Ravindran, B., & Cho, H. (2012). Probability-based prediction and sleep scheduling for energy-efficient target tracking in sensor networks. *IEEE Transactions on Mobile Computing, 12*(4), 735–747. doi:10.1109/TMC.2012.44

Mohamadi, H., Ismail, A. S., & Salleh, S. (2014). Solving target coverage problem using cover sets in wireless sensor networks based on learning automata. *Wireless Personal Communications, 75*(1), 447–463. doi:10.100711277-013-1371-x

Mulligan, R., & Ammari, H. M. (2010). Coverage in wireless sensor networks: a survey. *Network Protocols and Algorithms, 2*(2).

Thai, M.T., Wang, F., & Du, D-Z. (2008). Coverage problems in wireless sensor networks: designs and analysis. *International Journal of Sensor Networks, 3*(3), 191–200.

Wang, J., & Zhong, N. (2006). Efficient point coverage in wireless sensor networks. *Journal of Combinatorial Optimization, 11*(3), 291–304. doi:10.100710878-006-7909-z

Xu & Sahni. (2006). *Approximation Algorithms For Wireless Sensor Deployment*. Academic Press.

Yang, S., Wu, Y., & Xuan, J. (2007). *Time Series Analysis in Engineering Application*. HUST Press.

Ye, D., & Zhang, M. (2017). A self-adaptive sleep/wake-up scheduling approach for wireless sensor networks. *IEEE Transactions on Cybernetics, 48*(3), 979–992. doi:10.1109/TCYB.2017.2669996 PMID:28278488

Yick, J., Mukherjee, B., & Ghoshal, D. (2008). Wireless Sensor Networks Survey. *Computer Networks, 52*, 2292–2330.

Zhou, Z., Das, S., & Gupta, H. (2004). Connected k-coverage problem in sensor networks. *Proc. of International Conference on Computer Communications and Networks (ICCCN'04)*, 373–378.

Compilation of References

Sucasas, V., Mantas, G., & Rodriguez, J. (2016). Security challenges for cloud radio access networks. *Backhauling/Fronthauling for Future Wireless Systems*, 195-211.

Ullah, A., Sehr, I., Akbar, M., & Ning, H. (2018). *FoG assisted secure De-duplicated data dissemination in smart healthcare IoT. In 2018 IEEE international conference on smart internet of things (SmartIoT)*. IEEE. doi:10.1109/SmartIoT.2018.00038

Ahmad, I., Namal, S., Ylianttila, M., & Gurtov, A. (2015). Security in software defined networks: A survey. *IEEE Communications Surveys and Tutorials*, *17*(4), 2317–2346. doi:10.1109/COMST.2015.2474118

Butpheng, Yeh, & Xiong. (2020). Security and Privacy in IoT-Cloud-Based e-Health Systems—A Comprehensive Review. *Symmetry, 12*(7).

Fonseca, P., Bennesby, R., Mota, E., & Passito, A. (2012, April). *A replication component for resilient OpenFlow-based networking. In 2012 IEEE Network operations and management symposium*. IEEE.

Chaudhury, S., Paul, D., & Mukherjee, R. (2017). Haldar S Internet of Thing based healthcare monitoring system. In *Industrial automation and electromechanical engineering conference (IEMECON), 2017 8th annual* (pp. 346–349). IEEE. doi:10.1109/IEMECON.2017.8079620

Park, Y., & Park, T. (2007b, November). *A survey of security threats on 4G networks. In 2007 IEEE Globecom workshops*. IEEE.

El-hajj, M., Chamoun, M., Fadlallah, A., & Serhrouchni, A. (2017). Analysis of authentication techniques in internet of things (IoT). In *Cyber security in networking conference (CSNet), 1st* (pp. 1–3). IEEE. doi:10.1109/CSNET.2017.8242006

Strassner, J. (2004). Chapter 4-policy operation in a PBNM system. Policy-Based Network Management.

Aboushosha, B., Ramadan, R. A., Dwivedi, A. D., El-Sayed, A., & Dessouky, M. M. (2020). SLIM: A Lightweight Block Cipher for Internet of Health Things. *IEEE Access: Practical Innovations, Open Solutions, 8*, 203747–203757. doi:10.1109/ACCESS.2020.3036589

Twidle, K., Dulay, N., Lupu, E., & Sloman, M. (2009, April). Ponder2: A policy system for autonomous pervasive environments. In *2009 Fifth International Conference on Autonomic and Autonomous Systems* (pp. 330-335). IEEE. 10.1109/ICAS.2009.42

Fan, X., Mandal, K., & Gong, G. (2013). Wg-8: a lightweight stream cipher for resource-constrained smart devices. *Proceeding of International Conference on Heterogeneous Networking for Quality, Reliability.* 10.1007/978-3-642-37949-9_54

Zhou, J., Shen, Q., & Xu, Y. (2012, May). Research and improvement of Ponder2 policy language. In *2012 IEEE International Conference on Computer Science and Automation Engineering (CSAE)* (Vol. 2, pp. 455-458). IEEE. 10.1109/CSAE.2012.6272813

Li, L., Liu, B., & Wang, H. (2016). QTL: A new ultra-lightweight block cipher. *Microprocessors and Microsystems, 45*, 45–55. doi:10.1016/j.micpro.2016.03.011

Neisse, R., Costa, P. D., Wegdam, M., & van Sinderen, M. (2008, June). An information model and architecture for context-aware management domains. In *2008 IEEE Workshop on Policies for Distributed Systems and Networks* (pp. 162-169). IEEE. 10.1109/POLICY.2008.31

Biswas, K., Muthukkumarasamy, V., & Singh, K. (2015). An encryption scheme using chaotic map and genetic operations for wireless sensor networks. *IEEE Sensors Journal, 15*(5), 2801–2809. doi:10.1109/JSEN.2014.2380816

Zhao, H., Lobo, J., & Bellovin, S. M. (2008, June). An algebra for integration and analysis of ponder2 policies. In *2008 IEEE Workshop on Policies for Distributed Systems and Networks* (pp. 74-77). IEEE. 10.1109/POLICY.2008.42

Asim, M., Yautsiukhin, A., Brucker, A. D., Baker, T., Shi, Q., & Lempereur, B. (2018). Security policy monitoring of BPMN-based service compositions. *Journal of Software: Evolution and Process, 30*(9), e1944.

Tao, H., Bhuiyan, M. Z. A., Abdalla, A. N., Hassan, M. M., Zain, J. M., & Hayajneh, T. (2018). Secured data collection with hardware-based ciphers for IoT-based healthcare. *IEEE Internet of Things Journal,* 1–10.

Agiwal, M., Roy, A., & Saxena, N. (2016). Next generation 5G wireless networks: A comprehensive survey. *IEEE Communications Surveys and Tutorials, 18*(3), 1617–1655. doi:10.1109/COMST.2016.2532458

Thakor, V. A., Razzaque, M. A., & Khandaker, M. R. A. (2021). Lightweight Cryptography Algorithms for Resource-Constrained IoT Devices: A Review, Comparison and Research Opportunities. *IEEE Access: Practical Innovations, Open Solutions, 9*, 28177–28193. doi:10.1109/ACCESS.2021.3052867

Ismail, L., Materwala, H., & Zeadally, S. (2019). Lightweight Blockchain for Healthcare. *IEEE Access: Practical Innovations, Open Solutions, 7*, 149935–149951. doi:10.1109/ACCESS.2019.2947613

Töyssy, S., & Helenius, M. (2006). About malicious software in smartphones. *Journal in Computer Virology*, *2*(2), 109–119. doi:10.100711416-006-0022-0

Harikrishnan, T., & Babu, C. (2015). Cryptanalysis of hummingbird algorithm with improved security and throughput. *2015 International Conference on VLSI Systems, Architecture, Technology and Applications (VLSI-SATA)*, 1-6. 10.1109/VLSI-SATA.2015.7050460

Vook, F. W., Ghosh, A., & Thomas, T. A. (2014, June). MIMO and beamforming solutions for 5G technology. In *2014 IEEE MTT-S International Microwave Symposium (IMS2014)* (pp. 1-4). IEEE. 10.1109/MWSYM.2014.6848613

Goyal, S., Liu, P., Panwar, S. S., Difazio, R. A., Yang, R., & Bala, E. (2015). Full duplex cellular systems: Will doubling interference prevent doubling capacity? *IEEE Communications Magazine*, *53*(5), 121–127. doi:10.1109/MCOM.2015.7105650

Tan, C. C., Wang, H., Zhong, S., & Li, Q. (2009, November). IBE-Lite: A Lightweight Identity-Based Cryptography for Body Sensor Networks. *IEEE Transactions on Information Technology in Biomedicine*, *13*(6), 926–932. doi:10.1109/TITB.2009.2033055 PMID:19789117

Kim, H., Jung, I., Park, Y., Chung, W., Choi, S., & Hong, D. (2018). Time spread-windowed OFDM for spectral efficiency improvement. *IEEE Wireless Communications Letters*, *7*(5), 696–699. doi:10.1109/LWC.2018.2812150

Raza, A. R., Mahmood, K., Amjad, M. F., Abbas, H., & Afzal, M. (2020). On the efficiency of software implementations of lightweight block ciphers from the perspective of programming languages. *Future Generation Computer Systems*, *104*, 43–59. doi:10.1016/j.future.2019.09.058

Alassaf, N., Gutub, A., Parah, S. A., & Al Ghamdi, M. (2019). Enhancing speed of SIMON: A light-weight-cryptographic algorithm for IoT applications. *Multimedia Tools and Applications*, *78*(23), 32633–32657. doi:10.100711042-018-6801-z

Zheng, Y., He, D., Yu, W., & Tang, X. (2005, December). Trusted computing-based security architecture for 4G mobile networks. In *Sixth International Conference on Parallel and Distributed Computing Applications and Technologies (PDCAT'05)* (pp. 251-255). IEEE. 10.1109/PDCAT.2005.243

Park, Y., & Park, T. (2007a). A survey of security threats on 4G networks. *Proceedings of the IEEE Globecom Workshops*, 1–6. 10.1109/GLOCOMW.2007.4437813

Rodrigues, J. J. P. C., De Rezende Segundo, D. B., Junqueira, H. A., Sabino, M. H., Prince, R. M., Al-Muhtadi, J., & De Albuquerque, V. H. C. (2018). Enabling Technologies for the Internet of Health Things. *IEEE Access: Practical Innovations, Open Solutions*, *6*, 13129–13141. doi:10.1109/ACCESS.2017.2789329

Dierks, T., & Rescorla, E. (2008). *The transport layer security (TLS) protocol version 1.2*. Academic Press.

Khairuddin, A. M., Azir, K. N. F. K., & Kan, P. E. (2017). Limitations and future of electrocardiography devices: A review and the perspective from the Internet of Things. *International Conference on Research and Innovation in Information Systems*, 1-7. 10.1109/ICRIIS.2017.8002506

Santos, M. A., De Oliveira, B. T., Margi, C. B., Nunes, B. A., Turletti, T., & Obraczka, K. (2013, October). Software-defined networking based capacity sharing in hybrid networks. In *2013 21st IEEE international conference on network protocols (ICNP)* (pp. 1-6). IEEE. 10.1109/ICNP.2013.6733664

Xiao, L., Wan, X., Lu, X., Zhang, Y., & Wu, D. (2018). IoT Security Techniques Based on Machine Learning: How Do IoT Devices Use AI to Enhance Security? IEEE Signal Processing Magazine, 35(5), 41-49.

Nair, M. M., Tyagi, A. K., & Sreenath, N. (2021). The Future with Industry 4.0 at the Core of Society 5.0: Open Issues, Future Opportunities and Challenges. *2021 International Conference on Computer Communication and Informatics (ICCCI)*, 1-7. 10.1109/ICCCI50826.2021.9402498

YuHunag, C., MinChi, T., YaoTing, C., YuChieh, C., & YanRen, C. (2010, November). A novel design for future on-demand service and security. In *2010 IEEE 12th International Conference on Communication Technology* (pp. 385-388). IEEE.

Feng, W., Qin, Y., Zhao, S., & Feng, D. (2018). AAoT: Lightweight attestation and authentication of low-resource things in IoT and CPS. *Computer Networks, 134*, 167–182. doi:10.1016/j.comnet.2018.01.039

Namal, S., Ahmad, I., Gurtov, A., & Ylianttila, M. (2013, November). *Enabling secure mobility with OpenFlow. In 2013 IEEE SDN for Future Networks and Services (SDN4FNS)*. IEEE.

Alliance, N. G. M. N. (2015). 5G white paper. *Next generation mobile networks, white paper, 1*(2015).

Jafarian, J. H., Al-Shaer, E., & Duan, Q. (2012, August). Openflow random host mutation: transparent moving target defense using software defined networking. In *Proceedings of the first workshop on Hot topics in software defined networks* (pp. 127-132). 10.1145/2342441.2342467

Mohd, B. J., Hayajneh, T., & Vasilakos, A. V. (2015). A survey on lightweight block ciphers for low-resource devices: Comparative study and open issues. *Journal of Network and Computer Applications, 58*, 73–93. doi:10.1016/j.jnca.2015.09.001

Gember, A., Dragga, C., & Akella, A. (2012, October). ECOS: Leveraging software-defined networks to support mobile application offloading. In *2012 ACM/IEEE Symposium on Architectures for Networking and Communications Systems (ANCS)* (pp. 199-210). IEEE. 10.1145/2396556.2396598

Singh, S., Sharma, P. K., Moon, S. Y., & Park, J. H. (2017). Advanced lightweight encryption algorithms for IoT devices: Survey, challenges and solutions. *Journal of Ambient Intelligence and Humanized Computing, 4*, 1–18. doi:10.100712652-017-0494-4

Norrman, K., Näslund, M., & Dubrova, E. (2016, June). Protecting IMSI and user privacy in 5G networks. In *Proceedings of the 9th EAI international conference on mobile multimedia communications* (pp. 159-166). 10.4108/eai.18-6-2016.2264114

Stallings. (2017). *Cryptography and Network Security: Principles and Practice.* Academic Press.

Bhardwaj, I., Kumar, A., & Bansal, M. (2017). A review on lightweight cryptography algorithms for data security and authentication in IoTs. *Proc. 4th Int. Conf. Signal Process., Comput. Control (ISPCC),* 504–509. 10.1109/ISPCC.2017.8269731

3GPP. (2017). *SA3-Security. The Third Generation Partnership Project (3GPP).* Available: https://www.3gpp.org/Specifications-groups/sa-plenary/54-sa3-security

Khan, A. N., Kiah, M. M., Khan, S. U., & Madani, S. A. (2013). Towards secure mobile cloud computing: A survey. *Future Generation Computer Systems, 29*(5), 1278–1299. doi:10.1016/j.future.2012.08.003

Rashidi, B. (2020). Efficient and flexible hardware structures of the 128-bit CLEFIA block cipher. *IET Computers & Digital Techniques, 14*(2), 69–79. doi:10.1049/iet-cdt.2019.0157

Cai, Y., Yu, F. R., & Bu, S. (2015). Dynamic operations of cloud radio access networks (C-RAN) for mobile cloud computing systems. *IEEE Transactions on Vehicular Technology, 65*(3), 1536–1548.

Saravanan, P., Rani, S. S., Rekha, S. S., & Jatana, H. S. (2019). An Efficient ASIC Implementation of CLEFIA Encryption/Decryption Algorithm with Novel S-Box Architectures. *2019 IEEE 1st International Conference on Energy, Systems and Information Processing (ICESIP),* 1-6. 10.1109/ICESIP46348.2019.8938329

Jaeger, B. (2015, August). *Security orchestrator: Introducing a security orchestrator in the context of the etsi nfv reference architecture. In 2015 IEEE Trustcom/BigDataSE/ISPA* (Vol. 1). IEEE.

Jangra, M., & Singh, B. (2019). Performance analysis of CLEFIA and PRESENT lightweight block ciphers. *Journal of Discrete Mathematical Sciences and Cryptography, 22*(8), 1489–1499. doi:10.1080/09720529.2019.1695900

Ertaul, L., & Rajegowda, S. K. (2017). Performance Analysis of CLEFIA, PICCOLO, TWINE Lightweight Block Ciphers in IoT Environment. In *Proceedings of the International Conference on Security and Management (SAM). The Steering Committee of The World Congress in Computer Science, Computer Engineering and Applied Computing (WorldComp)* (pp. 25-31). Academic Press.

Lauer, H., & Kuntze, N. (2016, July). Hypervisor-based attestation of virtual environments. In *2016 Intl IEEE Conferences on Ubiquitous Intelligence & Computing, Advanced and Trusted Computing, Scalable Computing and Communications, Cloud and Big Data Computing, Internet of People, and Smart World Congress (UIC/ATC/ScalCom/CBDCom/IoP/SmartWorld)* (pp. 333-340). IEEE. 10.1109/UIC-ATC-ScalCom-CBDCom-IoP-SmartWorld.2016.0067

Bikos, A. N., & Sklavos, N. (2012). LTE/SAE security issues on 4G wireless networks. *IEEE Security and Privacy, 11*(2), 55–62. doi:10.1109/MSP.2012.136

Shirai, T., Shibutani, K., Akishita, T., Moriai, S., & Iwata, T. (2007, January). The 128-bit blockcipher CLEFIA. In *Fast software encryption* (pp. 181–195). Springer Berlin Heidelberg. doi:10.1007/978-3-540-74619-5_12

Tezcan, C. (2010). The improbable differential attack: Cryptanalysis of reduced round CLEFIA. In *Progress in Cryptology-INDOCRYPT 2010* (pp. 197-209). Springer Berlin Heidelberg.

Zhao, C., Huang, L., Zhao, Y., & Du, X. (2017). Secure machine-type communications toward LTE heterogeneous networks. *IEEE Wireless Communications*, *24*(1), 82–87. doi:10.1109/MWC.2017.1600141WC

Singh, S., Sharma, P. K., & Moon, S. Y. (2017). *Advanced lightweight encryption algorithms for IoT devices: survey, challenges and solutions. J Ambient Intell Human Comput.*

Ahmad, I., Liyanage, M., Ylianttila, M., & Gurtov, A. (2017, June). Analysis of deployment challenges of host identity protocol. In *2017 European Conference on Networks and Communications (EuCNC)* (pp. 1-6). IEEE. 10.1109/EuCNC.2017.7980675

Su, S., Dong, H., Fu, G., Zhang, C., & Zhang, M. (2014). A White-Box CLEFIA implementation for mobile devices. *2014 Communications Security Conference (CSC 2014)*, 1-8.

Mohd, B. J., Hayajneh, T., Ahmad Yousef, K. M., Khalaf, Z. A., & Bhuiyan, M. Z. A. (2018). Hardware design and modeling of lightweight block ciphers for secure communications. *Future Generation Computer Systems*, *83*, 83. doi:10.1016/j.future.2017.03.025

Andrews, J. G., Buzzi, S., Choi, W., Hanly, S. V., Lozano, A., Soong, A. C., & Zhang, J. C. (2014). What will 5G be? *IEEE Journal on Selected Areas in Communications*, *32*(6), 1065–1082. doi:10.1109/JSAC.2014.2328098

Qatan, F. M., & Damaj, I. W. (2012). High-speed KATAN ciphers on-a-chip. *2012 International Conference on Computer Systems and Industrial Informatics*, 1-6.

De Canniere, C., Dunkelman, O., & Knežević, M. (2009). KATAN and KTANTAN—A family of small and efficient hardware-oriented block ciphers. In *Proc. Int. Workshop Cryptograph. Hardw. Embedded Syst.* Springer. 10.1007/978-3-642-04138-9_20

Lu, L., Li, G. Y., Swindlehurst, A. L., Ashikhmin, A., & Zhang, R. (2014). An overview of massive MIMO: Benefits and challenges. *IEEE Journal of Selected Topics in Signal Processing*, *8*(5), 742–758. doi:10.1109/JSTSP.2014.2317671

Elijah, O., Leow, C. Y., Rahman, T. A., Nunoo, S., & Iliya, S. Z. (2015). A comprehensive survey of pilot contamination in massive MIMO—5G system. *IEEE Communications Surveys and Tutorials*, *18*(2), 905–923. doi:10.1109/COMST.2015.2504379

Mohd, B. J., Hayajneh, T., & Abu Khalaf, Z. (2015). Optimization and modeling of FPGA implementation of the KATAN Cipher. *2015 6th International Conference on Information and Communication Systems (ICICS)*, 68-72. 10.1109/IACS.2015.7103204

Abed, S., Jaffal, R., Mohd, B. J., & Alshayeji, M. (2019). FPGA Modeling and Optimization of a SIMON Lightweight Block Cipher. *Sensors (Basel)*, *19*(4), 913. doi:10.339019040913 PMID:30795605

Ahmad, I., Liyanage, M., Namal, S., Ylianttila, M., Gurtov, A., Eckert, M., . . . Ulas, A. (2016, January). New concepts for traffic, resource and mobility management in software-defined mobile networks. In *2016 12th Annual conference on wireless on-demand network systems and services (WONS)* (pp. 1-8). IEEE.

Costa-Requena, J., Santos, J. L., Guasch, V. F., Ahokas, K., Premsankar, G., Luukkainen, S., ... de Oca, E. M. (2015, June). SDN and NFV integration in generalized mobile network architecture. In *2015 European conference on networks and communications (EuCNC)* (pp. 154-158). IEEE. 10.1109/EuCNC.2015.7194059

AlKhzaimi, H., & Lauridsen, M. M. (2013). Cryptanalysis of the SIMON Family of Block Ciphers. *IACR Cryptol. ePrint Arch.*, 543.

Namal, S., Ahmad, I., Saud, S., Jokinen, M., & Gurtov, A. (2016). Implementation of OpenFlow based cognitive radio network architecture: SDN&R. *Wireless Networks*, *22*(2), 663–677. doi:10.100711276-015-0973-5

Ahmad, I. (2020). *Improving software defined cognitive and secure networking.* arXiv preprint arXiv:2007.05296.

AlAssaf, N., AlKazemi, B., & Gutub, A. (2017). Applicable Light-Weight Cryptography to Secure Medical Data in IoT Systems. *Journal of Research in Engineering and Applied Sciences*, *2*(2), 50–58. doi:10.46565/jreas.2017.v02i02.002

Kondo, K., Sasaki, Y., & Iwata, T. (2016). On the Design Rationale of Simon Block Cipher: Integral Attacks and Impossible Differential Attacks against Simon Variants. In M. Manulis, A. R. Sadeghi, & S. Schneider (Eds.), Lecture Notes in Computer Science: Vol. 9696. *Applied Cryptography and Network Security. ACNS 2016.* doi:10.1007/978-3-319-39555-5_28

Security, S. D. N. (2013). *Considerations in the Data Center.* Open Networking Foundation. Available: https://www.opennetworking.org/sdn-resources/sdn-library

Tawalbeh, Muheidat, Tawalbeh, & Quwaider. (2020). IoT Privacy and Security: Challenges and Solutions. *Applied Sciences, 10*(12).

Viega, M. M. J. (2003). *Secure Programming Cookbook for C and C++: Recipes for Cryptography, Authentication, Input Validation & More.* Academic Press.

Kölbl, S., Leander, G., & Tiessen, T. (2015). Observations on the SIMON Block Cipher Family. In R. Gennaro & M. Robshaw (Eds.), Lecture Notes in Computer Science: Vol. 9215. *Advances in Cryptology — CRYPTO 2015. CRYPTO 2015.* doi:10.1007/978-3-662-47989-6_8

Tao, H., Bhuiyan, M. Z. A., Abdalla, A. N., Hassan, M. M., Zain, J. M., & Hayajneh, T. (2019). Secured Data Collection with Hardware-Based Ciphers for IoT-Based Healthcare. *IEEE Internet of Things Journal*, *6*(1), 410–420. doi:10.1109/JIOT.2018.2854714

Dhanda, S. S., Singh, B., & Jindal, P. (2020). Lightweight Cryptography: A Solution to Secure IoT. *Wireless Personal Communications*, *112*(3), 1947–1980. doi:10.100711277-020-07134-3

Alladi, T., Chamola, V., Sikdar, B., & Choo, K.R. (2020). Consumer IoT: Security Vulnerability Case Studies and Solutions. *IEEE Consum. Electron.*, 17–25.

Chen, L., Thombre, S., Jarvinen, K., Lohan, E. S., Alen-Savikko, A., Leppakoski, H., Bhuiyan, M. Z. H., Bu-Pasha, S., Ferrara, G. N., Honkala, S., Lindqvist, J., Ruotsalainen, L., Korpisaari, P., & Kuusniemi, H. (2017). Robustness, Security and Privacy in Location-Based Services for Future IoT: A Survey. *IEEE Access: Practical Innovations, Open Solutions*, *5*, 8956–8977. doi:10.1109/ACCESS.2017.2695525

Gudeti, B., Mishra, S., Malik, S., Fernandez, T. F., Tyagi, A. K., & Kumari, S. (2020). A Novel Approach to Predict Chronic Kidney Disease using Machine Learning Algorithms. *2020 4th International Conference on Electronics, Communication and Aerospace Technology (ICECA)*, 1630-1635. 10.1109/ICECA49313.2020.9297392

Rekha, Reddy, & Tyagi. (2020). KDOS - Kernel Density based Over Sampling - A Solution to Skewed Class Distribution. *Journal of Information Assurance and Security, 15*(2), 44-52.

Rekha, G., Tyagi, A. K., & Krishna Reddy, V. (2019). Solving Class Imbalance Problem Using Bagging, Boosting Techniques, with and without Noise Filter Method. *International Journal of Hybrid Intelligent Systems*, *15*(2), 67–76. doi:10.3233/HIS-190261

Engineering, A. A. B. (2017). *Internet of Medical Things Revolutionizing Healthcare.* https://aabme.asme.org/posts/internet-of-medical-thingsrevolutionizing-healthcare

and the Future of Telemedicine and Remote Surgery I Digi International. (n.d.). Retrieved May 11, 2022, from https://www.digi.com/blog/post/5g-and-the-future-of-telemedicine-remote-surgery

Kulkarni, P., Khanai, R., & Bindagi, G. (2016, March). Security frameworks for mobile cloud computing: A survey. In 2016 international conference on electrical, electronics, and optimization techniques (ICEEOT) (pp. 2507-2511). IEEE. doi:10.1109/ICEEOT.2016.7755144

Nair, M. M., Kumari, S., Tyagi, A. K., & Sravanthi, K. (2021). Deep Learning for Medical Image Recognition: Open Issues and a Way to Forward. In *Proceedings of the Second International Conference on Information Management and Machine Intelligence. Lecture Notes in Networks and Systems* (vol. 166). Springer.

Pramod, A. (2020). *Machine Learning and Deep Learning: Open Issues and Future Research Directions for Next Ten Years. In Computational Analysis and Understanding of Deep Learning for Medical Care: Principles, Methods, and Applications, 2020.* Wiley Scrivener.

Tyagi. (2021, October). AARIN: Affordable, Accurate, Reliable and INnovative Mechanism to Protect a Medical Cyber-Physical System using Blockchain Technology. *IJIN, 2*, 175–183.

Madhav, A. V. S., & Tyagi, A. K. (2022). The World with Future Technologies (Post-COVID-19): Open Issues, Challenges, and the Road Ahead. In A. K. Tyagi, A. Abraham, & A. Kaklauskas (Eds.), *Intelligent Interactive Multimedia Systems for e-Healthcare Applications.* Springer. doi:10.1007/978-981-16-6542-4_22

Mishra, S., & Tyagi, A. K. (2022). The Role of Machine Learning Techniques in Internet of Things-Based Cloud Applications. In S. Pal, D. De, & R. Buyya (Eds.), *Artificial Intelligence-based Internet of Things Systems. Internet of Things (Technology, Communications and Computing).* Springer. doi:10.1007/978-3-030-87059-1_4

Nair, M. M., & Tyagi, A. K. (2021). Privacy: History, Statistics, Policy, Laws, Preservation and Threat Analysis. Journal of Information Assurance & Security, 16(1), 24-34.

Fuzon. (2019). *Internet of Medical Things (IoMT): New Era in Healthcare Industry.* Academic Press.

Vikas, S. S., Pawan, K., Gurudatt, A. K., & Shyam, G. (2014, February). Mobile cloud computing: Security threats. In 2014 international conference on electronics and communication systems (ICECS) (pp. 1-4). IEEE.

La Polla, M., Martinelli, F., & Sgandurra, D. (2012). A survey on security for mobile devices. *IEEE Communications Surveys and Tutorials, 15*(1), 446–471. doi:10.1109/SURV.2012.013012.00028

McKay, K., Bassham, L., Turan, M. S., & Mouha, N. (2017). *Report on Lightweight Cryptography (Nistir8114).* NIST.

Suo, H., Liu, Z., Wan, J., & Zhou, K. (2013, July). Security and privacy in mobile cloud computing. In *2013 9th International Wireless Communications and Mobile Computing Conference (IWCMC)* (pp. 655-659). IEEE. 10.1109/IWCMC.2013.6583635

Toshihiko, O. (2017). Lightweight cryptography applicable to various IoT devices. *NEC Tech. J., 12*(1), 67–71.

Biryukov, A., & Perrin, L. P. (2017). *State of the art in lightweight symmetric cryptography.* Univ. Luxembourg Library, Esch-sur-Alzette, Luxembourg, Tech. Rep. 10993/31319.

Chonka, A., & Abawajy, J. (2012, September). Detecting and mitigating HX-DoS attacks against cloud web services. In *2012 15th International Conference on Network-Based Information Systems* (pp. 429-434). IEEE. 10.1109/NBiS.2012.146

Aaronson, S. A. (2019). Data is different, and that's why the world needs a new approach to governing cross-border data flows. *Digital Policy, Regulation & Governance, 21*(5), 441–460. doi:10.1108/DPRG-03-2019-0021

Aazam, M., & Huh, E. N. (2014). Fog computing and smart gateway-based communication for cloud of things. *Proceedings of the 2nd IEEE International Conference on Future Internet of Things and Cloud (FiCloud' 14),* 464–470. 10.1109/FiCloud.2014.83

Abdualgalil, B., & Abraham, S. (2020). Applications of Machine Learning Algorithms and Performance Comparison: A Review. *IEEE 2020 International Conference on Emerging Trends in Information Technology and Engineering (ic-ETITE)*, 1–6. 10.1109/ic-ETITE47903.2020.490

Abdul-Ghani, H. A., Konstantas, D., & Mahyoub, M. (2018). A comprehensive IoT attacks survey based on a building-blocked reference model. *International Journal of Advanced Computer Science and Applications*, *9*(3), 355–373. doi:10.14569/IJACSA.2018.090349

Abduvaliyev, A., Pathan, A. S. K., Zhou, J., Roman, R., & Wong, W. C. (2013). On the vital areas of intrusion detection systems in wireless sensor networks. *IEEE Communications Surveys and Tutorials*, *15*(3), 1223–1237. doi:10.1109/SURV.2012.121912.00006

Abhishek, D., & Neha, G. (2021). A Systematic Review of Techniques, Tools and Applications of Machine Learning. *Proceedings of the Third International Conference on Intelligent Communication Technologies and Virtual Mobile Networks (ICICV 2021)*. 10.1109/ICICV50876.2021.9388637

Abomhara, M., & Koien, G. M. (2014). Security and privacy in the Internet of things: Current status and open issues. In *International Conference on Privacy and Security in Mobile System*. IEEE.

Abuladel, A., & Bamasag, O. (2020, March). Data and location privacy issues in IoT applications. In *2020 3rd International Conference on Computer Applications & Information Security (ICCAIS)* (pp. 1-6). IEEE.

Aggarwal, V. K. (2021). Integration of Blockchain and IoT (B-IoT): Architecture, Solutions, & Future Research Direction. In *IOP Conference Series: Materials Science and Engineering*. IOP Publishing.

Aggarwal. (2012). *RFID Security in the Context of "Internet of Things."* First International Conference on Security of Internet of Things, Kerala, India.

Ahad, A., Tahir, M., & Yau, K.-L. A. (2019). 5G-based smart healthcare network: Architecture, taxonomy, challenges and future research directions. *IEEE Access: Practical Innovations, Open Solutions*, *7*, 100747–100762. doi:10.1109/ACCESS.2019.2930628

Ahmad, I., Kumar, T., Liyanage, M., Okwuibe, J., Ylianttila, M., & Gurtov, A. (2018). Overview of 5G security challenges and solutions. *IEEE Communications Standards Magazine*, *2*(1), 36–43.

Akpakwu, G.A. (2017). A Survey on 5G Networks for the Internet of Things: Communication Technologies and Challenges. *IEEE Access*.

Akyildiz, I. F., Su, W., Sankarasubramaniam, Y., & Cayirci, E. (2002). Wireless Sensor Networks: A Survey. *Computer Networks*, *38*(4), 393–422. doi:10.1016/S1389-1286(01)00302-4

Alaba, F. A., Othman, M., Hashem, I. A. T., & Alotaibi, F. (2017). Internet of things security: A survey. *Journal of Network and Computer Applications*, *88*, 10–28. doi:10.1016/j.jnca.2017.04.002

Alamer, A. (2021). Security and privacy-awareness in a software-defined Fog computing network for the Internet of Things. *Optical Switching and Networking*, *41*, 100616.

Alam, S. (2020). Internet of things (IoT) enabling technologies, requirements, and security challenges. In *Advances in data and information sciences* (pp. 119–126). Springer.

Ali, S., Saad, W., & Rajatheya. (2020). *6G White Paper on Machine Learning in Wireless Communication Networks*. arXiv:2004.13875v1.

Ali-Shah, P. A., Habib, M., Sajjad, T., Umar, M., & Babar, M. (2017). Applications and Challenges Faced by Internet of Things - A Survey, in Lecture Notes of the Institute for Computer Sciences. *Social Informatics and Telecommunications Engineering Future Intelligent Vehicular Technologies, Springer, 2017*, 82–188.

Al-Karaki, J. N., & Gawanmeh, A. (2017). The optimal deployment, coverage, and connectivity problems in wireless sensor networks: Revisited. *IEEE Access: Practical Innovations, Open Solutions*, *5*, 8051–18065. doi:10.1109/ACCESS.2017.2740382

Alsamhi, S. H., Ma, O., Ansari, M. S., & Almalki, F. A. (2019). Survey on collaborative smart drones and internet of things for improving smartness of smart cities. *IEEE Access: Practical Innovations, Open Solutions*, *7*(September), 128125–128152. https://doi.org/10.1109/ACCESS.2019.2934998

Amanpreet, S., Narina, T., & Aksha, S. (2016). A review of supervised machine learning algorithms. *3rd International Conference on Computing for Sustainable Global Development (INDIACom)*.

Amruthnath, N., & Gupta, T. (2018). A research study on unsupervised machine learning algorithms for early fault detection in predictive maintenance. *5th International Conference on Industrial Engineering and Applications (ICIEA)*. 10.1109/IEA.2018.8387124

Anajemba, J. H., Yue, T., Iwendi, C., Chatterjee, P., Ngabo, D., & Alnumay, W. S. (2021). *A secure multi-user privacy technique for wireless IoT networks using stochastic privacy optimization*. IEEE Internet of Things Journal.

Anawar, M. R., Wang, S., Azam Zia, M., Jadoon, A. K., Akram, U., & Raza, S. (2018). Fog Computing: An Overview of Big IoT Data Analytics. *Wireless Communications and Mobile Computing*, *2018*, 1–22. Advance online publication. doi:10.1155/2018/7157192

Antonopoulos, A. M. (2014). *Mastering Bitcoin: Unlocking Digital Crypto-Currency* (1st ed.). O'Reilly Media, Inc.

Arduino. (2022). https://www.arduino.cc/en/main/arduinoBoardUno

Aslan, Ö. A., & Samet, R. (2020). A comprehensive review on malware detection approaches. *IEEE Access: Practical Innovations, Open Solutions*, *8*, 6249–6271. doi:10.1109/ACCESS.2019.2963724

Atlam, H. F., Alenezi, A., Alharthi, A., Walters, R. J., & Wills, G. B. (2017). Integration of cloud computing with Internet of Things: challenges and open issues. *Proceedings IEEE International Conference Internet of Things (iThings) and IEEE Green Computing and Communications (GreenCom) and IEEE Cyber, Physical Social Comput. (CPSCom) and IEEE Smart Data (SmartData)*, 670–675. doi:10.1109/iThings-GreenComCPSComSmartData.2017.105

Axelrod, C. W. (2015, May). Enforcing security, safety and privacy for the Internet of Things. In 2015 Long Island Systems, Applications and Technology (pp. 1-6). IEEE.

Axon, L. (2015). *Privacy-awareness in Blockchain-based PKI*. Technical Report.

Bahga, A. & Madisetti, V.K. (2016). *Blockchain platform for industrial Internet of Things*. Technical Report.

Bardyn, J., Melly, T., Seller, O., & Sornin, N. (2016). IoT: The era of LPWAN is starting now. In *Proceedings of the 42nd European Solid-State Circuits Conference, ESSCIRC Conference* (pp. 25–30). IEEE.

Baumgartner, M., Juhar, J., & Papaj, J. (2021). Short Performance Analysis of the LTE and 5G Access Technologies in NS-3. *Proceedings of the 16th Conference on Computer Science and Intelligence Systems, FedCSIS 2021, 25,* 337–340. doi:10.15439/2021F62

BLE. (n.d.). *Smart Bluetooth Low Energy*. http://www.bluetooth.com/Pages/Bluetooth-Smart.aspx

Blundo, C., Santis, A. D., Herzberg, A., Kutten, S., Vaccaro, U., & Yung, M. (1993). Perfectly-secure key distribution for dynamic conferences. In Lecture Notes in Computer Science: Vol. 740. *Advances in Cryptology* (pp. 471–486). Springer.

Bonomi, Milito, P. N., & J. Z. (2014). Fog Computing: A Platform for Internet of Things and Analytics. *Studies in Computational Intelligence*.

Bonomi, F., Milito, R., Natarajan, P., & Zhu, J. (2014). *Fog computing: a platform for internet of things and analytics. In Big Data and Internet of Things: A Road Map for Smart Environments*. Springer.

Bonomi, F., Milito, R., Zhu, J., & Addepalli, S. (2012). Fog computing and its role in the internet of things. *Proceedings of the 1st ACM MCC Workshop on Mobile Cloud Computing,* 13–16. 10.1145/2342509.2342513

Booth, G., Soknacki, A., & Somayaji, A. (2013). Cloud security: attacks and current defenses. *Proceedings 8th Annual Symposium on Information Assurance, ASIA13,* 4–5.

Borasia, S., & Raisinghani, V. (2011). A review of congestion control mechanisms for wireless sensor networks. In *Technology systems and management* (pp. 201–206). Springer. doi:10.1007/978-3-642-20209-4_28

Borgaonkar, R., Shaik, A., Asokan, N., Niemi, V. V., & Seifert, J.-P. (2015). *LTE and IMSI catcher myths*. Academic Press.

Borgia, E. (2014). The Internet of Things vision: Key features, applications, and open issues. *Computer Communications, 54,* 1–31. doi:10.1016/j.comcom.2014.09.008

Borgohain, T., Kumar, U., & Sanyal, S. (2015). Survey of Security and Privacy Issues of Internet of Things. *International Journal of Advanced Network Applications, 6*(4), 2372-2378.

Bouzarkouna, I., Sahnoun, M., Sghaier, N., Baudry, D., & Gout, C. (2018). Challenges Facing the Industrial Implementation of Fog Computing. *Proceedings - 2018 IEEE 6th International Conference on Future Internet of Things and Cloud, FiCloud 2018*, 341–348. 10.1109/FiCloud.2018.00056

Box, G., Jenkins, G., & Reinsel, G. (2008). *Time Series Analysis: Forecasting and Control*. Wiley.

Braeken, A., Liyanage, M., Kumar, P., & Murphy, J. (2019). Novel 5G authentication protocol to improve the resistance against active attacks and malicious serving networks. *IEEE Access: Practical Innovations, Open Solutions*, 7, 64040–64052. doi:10.1109/ACCESS.2019.2914941

Butun, I., Morgera, S. D., & Sankar, R. (2014). A survey of intrusion detection systems in wireless sensor networks. *IEEE Communications Surveys and Tutorials*, 16(1), 266–282. doi:10.1109/SURV.2013.050113.00191

Cardei, Thai, & Li. (2005). Energy-efficient target coverage in wireless sensor network. *Proc. Infocom'05*.

Challa, S., Wazid, M., Das, A. K., & Khan, M. K. (2018). Authentication protocols for implantable medical devices: Taxonomy, analysis and future directions. *IEEE Consum. Electron. Mag*, 7(1), 57–65.

Chan, H., Perrig, A., & Song, D. (2003). Random key pre distribution schemes for sensor networks. *Proc. 19th Int. Conf. Data Eng.*, 197–213.

Chaturvedi, P., & Daniel, A. K. (2015). An energy efficient node scheduling protocol for target coverage in wireless sensor networks. *5th International Conference on Communication System and Network Technologies(CSNT-2015)*.

Chen, Y., Liu, W., Niu, Z., Feng, Z., Hu, Q., & Jiang, T. (2020). Pervasive intelligent endogenous 6G wireless systems: Prospects, theories and key technologies. *Digital Communications and Networks, 6(3), 312–320. doi:10.1016/j.dcan.2020.07.002

Cheng, Y., & Agrawal, D. (2005). Efficient pairwise key establishment and management in static wireless sensor networks. *Proc. 2nd IEEE Int. Conf. Mobile Ad Hoc Sensor Syst.*, 7.

Chen, H., Meng, C., Shan, Z., Fu, Z., & Bhargava, B. K. (2019). A novel low-rate denial of service attack detection approach in ZigBee wireless sensor network by combining Hilbert–Huang transformation and trust evaluation. *IEEE Access: Practical Innovations, Open Solutions*, 7, 32853–32866.

Chen, M., Hao, Y., Li, Y., Lai, C.-F., & Wu, D. (2015, June). On the computation offloading at ad hoc cloudlet: Architecture and service modes. *IEEE Communications Magazine*, 53(6), 1824.

Chen, M., Qian, Y., Mao, S., Tang, W., & Yang, X. (2016). Software-Defined Mobile Networks Security. *Mobile Networks and Applications*, 21(5), 729–743. doi:10.100711036-015-0665-5

Cheruvu, S. (2020). Connectivity technologies for IoT. In *Demystifying internet of things security* (pp. 347–411). Apress.

Chowdhury, S., & Schoen, M. P. (2020). *Classification using Supervised Machine Learning Techniques. Intermountain Engineering Technology and Computing.* doi:10.1109/ietc47856.2020.924921

Christidis, K., & Devetsikiotis, M. (2016). Blockchains and Smart contracts for the Internet of Things. *IEEE Access: Practical Innovations, Open Solutions, 4*, 2292–2303. doi:10.1109/ACCESS.2016.2566339

Cirani, S., Ferrari, G., & Veltri, L. (2013). Enforcing security mechanisms in the IP-based internet of things: An algorithmic overview. *Algorithms, 6*(2), 197–226. doi:10.3390/a6020197

Clincy, V., & Shahriar, H. (2019). IoT malware analysis. *Proc. IEEE 43rd Annu. Comput. Softw. Appl. Conf. (COMPSAC), 1*, 920–921.

Columbus, L. (2016). *Roundup Of Internet Of Things Forecasts And Market Estimates.* Academic Press.

Conoscenti, M., Vetro, A., & Martin, J. C. D. (2016). Blockchain for the Internet of Things: A systematic literature Review. *3rd International Symposium on Internet of Things: Systems, Management, and Security, IOTSMS-2016.*

Contiki-OS. (2022). http://www.contiki-os.org/

Costanzo & Masotti. (2017). Energizing 5G. *IEEE Microwave Magazine.*

Craciunescu, M., Chenaru, O., Dobrescu, R., Florea, G., & Mocanu, S. (2020). *IIoT Gateway for Edge Computing Applications.* . doi:10.1007/978-3-030-27477-1_17

Dalal, R., & Kushal, R. (2018). Review on Application of Machine Learning Algorithm for Data Science. *IEEE 3rd International Conference on Inventive Computation Technologies (ICICT),* 270-273. 10.1109/ICICT43934.2018.9034256

Das, A. K., & Zeadally, S. (2019). Data security in the smart grid environment. In Pathways to a Smarter Power System. Academic.

De Donno, M., Tange, K., & Dragoni, N. (2019). Foundations and Evolution of Modern Computing Paradigms: Cloud, IoT, Edge, and Fog. *IEEE Access: Practical Innovations, Open Solutions, 7*, 150936–150948. doi:10.1109/ACCESS.2019.2947652

De Hert, P., Papakonstantinou, V., Malgieri, G., Beslay, L., & Sanchez, I. (2018). The right to data portability in the GDPR: Towards user-centric interoperability of digital services. *Computer Law & Security Review, 34*(2), 193–203. doi:10.1016/j.clsr.2017.10.003

Deogirikar, J. (2017). Security Attacks inIoT. *Survey (London, England),* 32–37.

Deshmukh, A., Sreenath, N., Tyagi, A. K., & Abhichandan, U. V. E. (2022, January). Blockchain Enabled Cyber Security: A Comprehensive Survey. In *2022 International Conference on Computer Communication and Informatics (ICCCI)* (pp. 1-6). IEEE. 10.1109/ICCCI54379.2022.9740843

Devices & Systems. IoT Tech Expo. (2019). *Unlocking IoT Data With 5G and AI*. Available: https://innovate.ieee.org/innovation-spotlight/5g-iot-ai/

Dinu, D., Biryukov, A., & Großschädl, J. (2015). FELICS – fair evaluation of lightweight cryptographic systems. In *NIST Workshop on Lightweight Cryptography*. National Institute of Standards and Technology (NIST).

Dongand, Q., & Liu, D. (2007). Using auxiliary sensors for pairwise key establishment in WSN. *Proc. IFIP Int. Conf. Netw. (Networking), 251–262*.

Du, W., Deng, J., Han, Y. S., Chen, S., & Varshney, P. K. (2004). A key management scheme for wireless sensor networks using deployment knowledge. *Proc. 23rd Conf. IEEE Commun. Soc. (Infocom), 1*, 586–597.

Du, J., & Jiang, C. (2020). Machine Learning for 6G Wireless Networks: Carry-Forward-Enhanced Bandwidth, Massive Access, and Ultrareliable/Low Latency. *IEEE Vehicular Technology Magazine*. Advance online publication. doi:10.1109/MVT.2020.3019650

Du, W., Deng, J., Han, Y. S., & Varshney, P. K. (2003). A pairwise key pre distribution scheme for wireless sensor networks. *Proc. 10th ACM Conf. Comput. Commun. Secur.(CCS), 42–51*.

Elazhary, H. (2019). Internet of Things (IoT), mobile cloud, cloudlet, mobile IoT, IoT cloud, fog, mobile edge, and edge emerging computing paradigms: Disambiguation and research directions. *Journal of Network and Computer Applications, 128*(October), 105–140. doi:10.1016/j. jnca.2018.10.021

Ericsson. (2019). *What is 5G?* Available: https://www.ericsson.com/en/5g

Eschenauer, L., & Gligor, V. D. (2002). A key management scheme for distributed sensor networks. *Proc. 9th ACM Conf. Comput. Commun. Secur., 41–47*.

Fernandez, V., Gordon, K. J., Collins, R. J., Townsend, P. D., Cova, S. D., Rech, I., & Buller, G. S. (2005, July). Quantum key distribution in a multi-user network at gigahertz clock rates. In Photonic Materials, Devices, and Applications (Vol. 5840, pp. 720-727). International Society for Optics and Photonics.

Friese, I., Heuer, J., & Kong, N. (2014). Challenges from the Identities of Things: Introduction of the definitive of Things discussion group with Kantara initiative. *IEEE World Forum on Internet of Things (WE-IoT), 1-4*.

Fromknecht, C., Velicanu, D., & Yakoubov, S. (2014). *CertCoin: A namecoin based decentralized authentication system*. Academic Press.

Fuentes, M., & Carcel, J. L. (2018, June). 5G New Radio Evaluation against IMT-2020 Key Performance Indicators. *IEEE Access: Practical Innovations, Open Solutions, 8*, 110880–110896. https://doi.org/10.1109/ACCESS.2020.3001641

Gandotra & Jha. (2017). A survey on green communication and security challenges in 5G wireless communication networks. *Journal of Network and Computer Applications, 96*, 39–61.

Gautam, S., Malik, A., Singh, N., & Kumar, S. (2019). Recent Advances and Countermeasures against Various Attacks in IoT Environment. *2nd International Conference on Signal Processing and Communication, ICSPC 2019 - Proceedings*, 315–319. 10.1109/ICSPC46172.2019.8976527

Github. (2022). *WLAN Overview (Roaming- current and future enhancements).* http://what-when-how.com/roaming-in-wireless-networks/wlan-overview-roamingcurrent-and-future-enhancements/

Gomathi, R. M., Krishna, G. H. S., Brumancia, E., & Dhas, Y. M. (2018). A Survey on IoT Technologies, Evolution and Architecture. *2nd International Conference on Computer, Communication, and Signal Processing: Special Focus on Technology and Innovation for Smart Environment, ICCCSP 2018,* 1–5. 10.1109/ICCCSP.2018.8452820

Goyat, R., Kumar, G., Saha, R., Conti, M., Rai, M. K., Thomas, R., & Hoon-Kim, T. (2020). *Blockchain-based data storage with privacy and authentication in internet-of-things.* IEEE Internet of Things Journal.

Granjal, Monteiro, & Sa Silva. (2015). Security for the Internet of Things: A survey of existing protocols and open research issues. *IEEE Commun. Surveys Tuts., 17*(3), 1294–1312.

Granjal, J., Monteiro, E., & Silva, J. S. (2015). Security for the internet of things: A Survey of existing protocols and open research issues. *IEE Communication Surveys and Tutorial, 17*(3), 1294–1312. doi:10.1109/COMST.2015.2388550

Granjal, J., Silva, R., Monteriro, E., Silva, J. S., & Boavida, F. (2008). Why is IPSec a viable option for wireless sensor networks. *5th IEEE International Conference on Mobile Ad Hoc and Sensor Systems*, 802-807. 10.1109/MAHSS.2008.4660130

Gubbi, J., Buyya, R., Marusic, S., & Palaniswami, M. (2013). Internet of Things (IoT): A vision, architectural elements, and future directions. *Future Generation Computer Systems, 29*(7), 1645–1660. doi:10.1016/j.future.2013.01.010

Guo, W. (2019). *Explainable artificial intelligence (XAI) for 6G: Improving trust between human and machine.* arXiv preprint arXiv:1911.04542.

Gupta, S. S., Khan, M. S., & Sethi, T. (2019, June). Latest trends in security, privacy and trust in IOT. In *2019 3rd International conference on Electronics, Communication and Aerospace Technology (ICECA)* (pp. 382-385). IEEE.

Haenlein, M., & Kaplan, A. (2019). A brief history of artificial intelligence: On the past, present, and future of artificial intelligence. *California Management Review, 61*(4), 5–14. doi:10.1177/0008125619864925

Hakak, S. A., Latif, S., & Amin, G. (2013). A Review on Mobile Cloud Computing and Issues in it. *International Journal of Computers and Applications, 75*(11), 1–4. doi:10.5120/13152-0760

Hameed, A., & Alomary, A. (2019, September). Security issues in IoT: a survey. In *2019 International conference on innovation and intelligence for informatics, computing, and technologies (3ICT)* (pp. 1-5). IEEE.

He, D., Kumar, N., Khan, M. K., Wang, L., & Shen, J. (2018). Efficient privacy aware authentication scheme for mobile cloud computing services. *IEEE Systems Journal, 12*(2), 1621–1631.

Heinzelman, W. R., Chandrakasan, A., & Balakrishnan, H. (2000). Energy-efficient communication protocol for wireless micro sensor networks. In *HICSS '00: Proceedings of the 33rd Hawaii International Conference on System Sciences (vol. 9).* IEEE Computer Society.

Hewa, T., Gurkan, G., Kalla, A., & Ylianttila. (2020). The Role of Blockchain in 6G: Challenges, Opportunities and Research Directions. *2nd 6G wireless Summit (6G SUMMIT).* . doi:10.1109/6 GSUMMIT49458.2020.9083784

Hong, X., Jing, W., & Jianghong, S. (2014). Cognitive radio in 5G: A perspective on energy-spectral efficiency trade–off. *IEEE Communications Magazine, 52*(7), 46–53. Advance online publication. doi:10.1109/MCOM.2014.6852082

Hou, J., Qu, L., & Shi, W. (2019). A survey on Internet of Things security from data perspectives. *Computer Networks, 48*, 295–306.

How to Avoid the Dreaded Computer Virus. (n.d.). Available: http://www.magellansolutions. co.uk/malware.html

Hsu, T. C., Yang, H., Chung, Y. C., & Hsu, C. H. (2020). A Creative IoT agriculture platform for cloud fog computing. *Sustainable Computing: Informatics and Systems, 28*, 100285. doi:10.1016/j. suscom.2018.10.006

https://medium.com/illumination/5g-fifth-generation-of-mobile-networks-part-1f32d7f003686

Hu, C., Zhang, J., & Wen, Q. (2011, October). An identity-based personal location system with protected privacy in IoT. In *2011 4th IEEE International Conference on Broadband Network and Multimedia Technology* (pp. 192-195). IEEE.

Hussain, R., & Zeadally, S. (2019). Autonomous cars: Research results, issues, and future challenges, *IEEE Commun. Surveys Tuts., 21*(2), 1275–1313.

IEEE. (2012). https://ieeexplore.ieee.org/document/6185525

Imran, M. A., Sambo, Y. A., & Abbasi, Q. H. (2019). Evolution of vehicular communications within the context of 5G systems. In *Enabling 5G Communication Systems to Support Vertical Industries* (pp. 103–126). IEEE. doi:10.1002/9781119515579.ch5

Internet of Things–Architecture. (2013). *IoT-A Deliverable D1.5 –Final architectural reference model for the IoT v3.0.* Academic Press.

Iqbal, H. S. (2021). Machine Learning: Algorithms, Real World Applications and Research Directions. *SN Computer Science, 2*(160). doi:10.1007/s42979-021-00592-x

Jaggi, N., & Abouzeid, A. A. (2006). Energy-Efficient Connected Coverage in Wireless Sensor Networks. *Proc. 4th Asian International Mobile Computing Conference (AMOC)*, 77-86.

Jan, S. U., Ahmed, S., Shakhov, V., & Koo, I. (2019). Toward a lightweight intrusion detection system for the Internet of Things. *IEEE Access: Practical Innovations, Open Solutions, 7,* 42450–42471.

Jiang, B., Ravindran, B., & Cho, H. (2012). Probability-based prediction and sleep scheduling for energy-efficient target tracking in sensor networks. *IEEE Transactions on Mobile Computing, 12*(4), 735–747. doi:10.1109/TMC.2012.44

Jo, K., & Sunwoo, M. (2014). Generation of a precise roadway map for autonomous cars. *IEEE Trans. Intell. Transp., 15*(3), 925–937.

Kamaldeep, M. M., & Dutta, M. (2017). Contiki-based mitigation of UDP f looding attacks in the Internet of Things. *Proc. Int. Conf. Comput., Commun. Automat. (ICCCA),* 1296–1300. doi: 10.1109/CCAA.2017.8229997

Kao, Y.-W., Huang, K.-Y., Gu, H.-Z., & Yuan, S.-M. (2013). UCloud: A usercentric key management scheme for cloud data protection. *IET Information Security, 7*(2), 144–154.

Kaul, V., Enslin, S., & Gross, S. A. (2020). History of artificial intelligence in medicine. *Gastrointestinal Endoscopy, 92*(4), 807–812. doi:10.1016/j.gie.2020.06.040 PMID:32565184

Kerckhof, S., Durvaux, F., Hocquet, C., Bol, D., & Standaert, F. X. (2012). *Towards green cryptography: a comparison of lightweight ciphers from the energy viewpoint. In Cryptographic hardware and embedded systems–CHES 2012.* Springer.

Khan, M., Quasim, M., Algarni, F., & Alharthi, A. (2020). *Decentralised Internet of Things: A Blockchain Perspective.* doi:10.1007/978-3-030-38677-1

Khan, M. A., & Salah, K. (2018). IoT security: Review, block chain solutions, and open challenges. *Future Generation Computer Systems, 82,* 395–411.

Khanna, A., & Kaur, S. (2019). Evolution of Internet of Things (IoT) and its significant impact in the field of Precision Agriculture. *Computers and Electronics in Agriculture, 157*(January), 218–231. doi:10.1016/j.compag.2018.12.039

Khan, R., Kumar, P., Jayakody, D. N. K., & Liyanage, M. (2020). A survey on security and privacy of 5G technologies: Potential solutions, recent advancements, and future directions. *IEEE Commun., 22*(1), 196–248. doi:10.1109/COMST.2019.2933899

Khan, W. Z., Ahmed, E., Hakak, S., Yaqoob, I., & Ahmed, A. (2019). Edge computing: A survey. *Future Generation Computer Systems, 97,* 219–235. doi:10.1016/j.future.2019.02.050

Kien, T., Long Ton, T., & Nguyen, G. M. T. (2020). Plant Leaf Disease Identification by Deep Convolutional Autoencoder as a Feature Extraction Approach. *Proceedings of The IEEE 17ᵗʰ International Conference on Electrical Engineering / Electronics, Computer, Telecommunications and Information Technology (ECTI-CON).* Available on https://ieeexplore. ieee.org/document/9158218/

Kiran, S., & Sriramoju, S. B. (2018). A study on the applications of IOT. *Indian Journal of Public Health Research & Development, 9*(11), 1173. https://doi.org/10.5958/0976-5506.2018.01616.9

Kshetri, N. (2017). Can blockchain strengthen the Internet of Things? *IT Professional, 19*(4), 68–72.

Kumar, R., Zhang, X., Wang, W., Khan, R. U., Kumar, J., & Sharif, A. (2019). A multimodal malware detection technique for Android IoT devices using various features. *IEEE Access: Practical Innovations, Open Solutions, 7*, 64411–64430.

Law, Y., Doumen, J., & Hartel, P. (2006). Survey and benchmark of block ciphers for wireless sensor networks. *ACM Transactions on Sensor Networks, 2*(1), 65–93. doi:10.1145/1138127.1138130

Lee, H. (2015). *Concept and Characteristics of 5G Mobile Communication Systems.* Available: https://www.netmanias.com/en/post/blog/7109/5g-iot/concept-andcharacteristics-of-5g-mobile-communication-systems-1

Liao, D., Li, H., Sun, G., Zhang, M., & Chang, V. (2018). Location and trajectory privacy preservation in 5G-Enabled vehicle social network services. *Journal of Network and Computer Applications, 110*, 108–118. doi:10.1016/j.jnca.2018.02.002

Liao, S. K., Cai, W. Q., Liu, W. Y., Zhang, L., Li, Y., Ren, J. G., ... Pan, J. W. (2017). Satellite-to-ground quantum key distribution. *Nature, 549*(7670), 43–47.

Li, J., Chen, X., Li, M., Li, J., Lee, P. P. C., & Lou, W. (2014). Secure deduplication with efficient and reliable convergent key management. *IEEE Transactions on Parallel and Distributed Systems, 25*(6), 1615–1625.

Limitless Intelligent ConnectivityM. W. C. (n.d.). Available: https://www.mwcbarcelona.com/

Li, S., Da Xu, L., & Zhao, S. (2018). 5G Internet of Things: A survey. *Journal of Industrial Information Integration, 10*, 1–9. doi:10.1016/j.jii.2018.01.005

Li, T., Sahu, A. K., Talwalkar, A., & Smith, V. (2020). Federated Learning: Challenges, Methods, and Future Directions. *IEEE Signal Processing Magazine, 37*(3), 50–60. doi:10.1109/MSP.2020.2975749

Liu, D., Ning, P., & Du, W. (n.d.). Group-based key pre-distribution in wireless sensor networks. *Proc. ACM Workshop Wireless Secur. (WiSe),* 1–14.

Liu, D., Ning, P., & Li, R. (2005). Establishing pairwise keys in distributed sensor networks. *ACM Transactions on Information and System Security, 8*(1), 41–77.

Liu, Z., Zhang, L., Ni, Q., Chen, J., Wang, R., Li, Y., & He, Y. (2019). An integrated architecture for IoT malware analysis and detection. In B. Li, M. Yang, H. Yuan, & Z. Yan (Eds.), *IoT as a Service* (pp. 127–137). Springer.

Li, Y., & Yang, R. (2021). Edge Computing Offloading Strategy Based on Dynamic Non-cooperative Games in D-IoT. *Wireless Personal Communications.* Advance online publication. doi:10.100711277-021-08891-5

Liyanage, M., Salo, J., Braeken, A., Kumar, T., Seneviratne, S., & Ylianttila, M. (2018). 5G Privacy: Scenarios and Solutions. *IEEE 5G World Forum, 5GWF 2018 - Conference Proceedings*, 197–203. 10.1109/5GWF.2018.8516981

Mahalle, P., Babar, S., Prasad, N. R., & Prasad, R. (2010). Identity Management Framework towards Internet of Things (IoT): Roadmap and Key Challenges. In *Recent trends in network security and applications, communications in computer and information science*. Springer.

Marmol, C., Pablo, F. S., Aurora, G. V., Jose, L., Hernandez, R., Jorge, B., Gianmarco, B., & Antonio, S. (2021). *Evaluating Federated Learning for Intrusion Detection in Internet of Things: Review and Challenges Enrique*. arXiv:2108.00974v1.

Messerges, T. S., Dabbish, E. A., & Sloan, R. H. (2002). Examining smartcard security under the threat of power analysis attacks. *IEEE Transactions on Computers*, *51*(5), 541–552.

Mingzhe, C., Deniz, G., Kaibin, H., Walid, S., & Mehdi, B. (2021). Distributed Learning in Wireless Networks: Recent Progress and Future Challenges. *IEEE Journal on Selected Areas in Communications*, *39*(12), 3579–3605. doi:10.1109/JSAC.2021.3118346

Mishra, S., & Tyagi, A. K. (2019, December). Intrusion detection in Internet of Things (IoTs) based applications using blockchain technolgy. In *2019 Third International conference on I-SMAC (IoT in Social, Mobile, Analytics and Cloud)(I-SMAC)* (pp. 123-128). IEEE. 10.1109/I-SMAC47947.2019.9032557

Mitchell, R., & Chen, I. R. (2014). Review: A survey of intrusion detection in wireless network applications. *Computer Communications*, *42*, 1–23. doi:10.1016/j.comcom.2014.01.012

Mohamadi, H., Ismail, A. S., & Salleh, S. (2014). Solving target coverage problem using cover sets in wireless sensor networks based on learning automata. *Wireless Personal Communications*, *75*(1), 447–463. doi:10.100711277-013-1371-x

Mohanta, B. K., Jena, D., Ramasubbareddy, S., Daneshmand, M., & Gandomi, A. H. (2020). Addressing security and privacy issues of IoT using blockchain technology. *IEEE Internet of Things Journal*, *8*(2), 881–888. doi:10.1109/JIOT.2020.3008906

Mollah, M. B., Zeadally, S., & Azad, M. A. K. (2019). *Emerging wireless technologies for Internet of Things applications: Opportunities and challenges. In Encyclopaedia of Wireless Networks*. Springer International Publishing Cham.

Mulligan, R., & Ammari, H. M. (2010). Coverage in wireless sensor networks: a survey. *Network Protocols and Algorithms*, *2*(2).

Nakamoto, S. (2008). *Bitcoin: A peer-to-peer electronic cash system*. Academic Press.

Nawir, M., Amir, A., Yaakob, N., Lynn, O. B., & Engineering, C. (2011). 2014 2nd International Conference on Electronic Design, ICED 2014. *2014 2nd International Conference on Electronic Design, ICED 2014*.

Compilation of References

Nawir, M., Amir, A., Yaakob, N., & Lynn, O. B. (2016). Internet of Things (IoT): Taxonomy of security attacks. *3rd International Conference on Electronic Design*, 321-326. 10.1109/ICED.2016.7804660

Nodered. (2022). https://nodered.org/

Nokia. (2016). *LTE Evolution for IoT Connectivity.* Nokia, Tech. Rep.

Northeast Now. China: Shanghai's Hongkou District Becomes First With 5G Network in World. (n.d.). Available: https://nenow.in/neighbour/china-shanghais-hongkou-district-becomesfirst-with-5g-network-in-world.html

Novo, O. (2018). Blockchain Meets IoT. *An Architecture for Scalable Access Management in IoT, 5*, 1184–1195. doi:10.1109/JIOT.2018.2812239

Oleshchuk, V. (2009). Internet of things and privacy preserving technologies. *2009 1st International Conference on Wireless Communication, Vehicular Technology, Information Theory and Aerospace Electronic Systems Technology*, 336–340.

Otte, P., de Vos, M., & Pouwelse, J. (2017). TrustChain: A Sybil-resistant scalable blockchain. *Future Generation Computer Systems.*

Pal, K. (2019). Algorithmic Solutions for RFID Tag Anti-Collision Problem in Supply Chain Management. *Procedia Computer Science*, 929-934.

Pal, K. (2021a). Privacy, Security and Policies: A Review of Problems and Solutions with Blockchain-Based Internet of Things Applications in Industrial Industry. *Procedia Computer Science.*

Pal, K. (2021b). A Novel Frame-Slotted ALOHA Algorithm for Radio Frequency Identification System in Supply Chain Management. *Procedia Computer Science*, 871-876. 10.1016/j.procs.2021.03.110

Pal, K. (2022a). Application of Game Theory in Blockchain-Based Healthcare Information System. In Prospects of Blockchain Technology for Accelerating Scientific Advancement in Healthcare. The IGI Global Publishing.

Pal, K. (2022b). A Decentralized Privacy Preserving Healthcare Blockchain for IoT, Challenges and Solutions. In Prospects of Blockchain Technology for Accelerating Scientific Advancement in Healthcare. The IGI Global Publishing.

Pal, K. (2022b). Semantic Interoperability in Internet of Things: Architecture, Protocols, and Research Challenges. In Management Strategies for Sustainability, New Knowledge Innovation, and Personalized Products and Services. The IGI Global Publishing.

Pal, K., & Yasar, A. (2020b). Semantic Approach to Data Integration for an Internet of Things Supporting Apparel Supply Chain Management. *Procedia Computer Science*, 197 - 204.

Pal, K., & Yasar, K. (2020a). Internet of Things and Blockchain Technology in Apparel Manufacturing Supply Chain Data Management. *Procedia Computer Science*, 450 - 457.

Perrig, A., Szewczyk, R., Tygar, J. D., Wen, V., & Culler, D. E. (2002). SPINS: Security protocols for sensor networks. *Wireless Networks*, 8(5), 521–534.

Pham, C., Nguyen, T., & Dang, T. (2019). *Resource-Constrained IoT Authentication Protocol: An ECC-Based Hybrid Scheme for Device-to-Server and Device-to-Device Communications.* doi:10.1007/978-3-030-35653-8_30

Phommasan, B., Jiang, Z., & Zhou, T. (2019, September). Research on Internet of Things Privacy Security and Coping Strategies. In *2019 International Conference on Virtual Reality and Intelligent Systems (ICVRIS)* (pp. 465-468). IEEE.

Pongle, P., & Chavan, G. (2015). A survey: Attacks on RPL and 6LoWPAN in IoT. *2015 International Conference on Pervasive Computing: Advance Communication Technology and Application for Society, ICPC 2015.* 10.1109/PERVASIVE.2015.7087034

Pureswaran, V., & Brody, P. (2014). *Device Democracy – Saving the future of the Internet of Things.* IBM.

Qualcomm. (2019). *Everything You Need to Know About 5G.* Available: https://www.qualcomm.com/invention/5

Raspberrypi. (2022). https://www.raspberrypi.org

Rass, S., Konig, S., & Schauer, S. (2015). BB84 quantum key distribution with intrinsic authentication. *ICQNM*, 41-44.

Rekha, G., Malik, S., Tyagi, A. K., & Nair, M. M. (2020). Intrusion detection in cyber security: Role of machine learning and data mining in cyber security. *Advances in Science. Technology and Engineering Systems Journal*, 5(3), 72–81. doi:10.25046/aj050310

Roman, R., Lopez, J., & Mambo, M. (2016). Mobile edge computing. In *A survey and analysis of security threats and challenges.* Future Generation Computer Systems.

Roman, R., Alcaraz, C., Lopez, J., & Sklavos, N. (2011). Key management systems for sensor networks in the context of the internet of things. *Computers & Electrical Engineering*, 37(2), 147–159. doi:10.1016/j.compeleceng.2011.01.009

RPMA. (2016). *RPMA Technology for the Internet of Things.* Ingenu, Tech. Rep.

Rupali, Shinde, & Yogita. (2020). Internet of Things Security Risk and Challenges. *IRJET*.

Said, O. & Masud, M. (2013). Towards Internet of Things: Survey and Future Vision. *International Journal of Computer Networks, 5*(1), 1–17.

Salh, A., Audah, L., & Shah, N. S. M. (2020). A Survey on Deep Learning for Ultra-Reliable and Low-Latency Communications Challenges on 6G Wireless Systems. *Proceedings of Future of Information and Communication Conference (FICC) 2021.* arXiv: 2004.08549v3.

Schurgot, M. R., Shinberg, D. A., & Greenwald, L. G. (2015, June). Experiments with security and privacy in IoT networks. In *2015 IEEE 16th International Symposium on a World of Wireless, Mobile and Multimedia Networks (WoWMoM)* (pp. 1-6). IEEE.

Shafi, M., Molisch, A. F., Smith, P. J., Haustein, T., Zhu, P., De Silva, P., Tufvesson, F., Benjebbour, A., & Wunder, G. (2017). 5G: A tutorial overview of standards, trials, challenges, deployment, and practice. *IEEE Journal on Selected Areas in Communications, 35*(6), 1201–1221.

Shannon, C. E. (1949). Communication Theory of Secrecy Systems. *The Bell System Technical Journal, 28*(4), 656–715. doi:10.1002/j.1538-7305.1949.tb00928.x

Sheena, A., & Sachin, A. (2017). Machine learning and its applications: A review. *International Conference on Big Data Analytics and Computational Intelligence (ICBDAC).* 10.1109/ICBDACI.2017.8070809

Shelby, K., Hartke, K., & Bormann, C. (2014). *The constrained application protocol (CoAP).* https://tools.ietf.org/html/rfc7252

Shi, W., Cao, J., Zhang, Q., Li, Y., & Xu, L. (2016). Edge Computing: Vision and Challenges. *IEEE Internet of Things Journal, 3*(5), 637–646. doi:10.1109/JIOT.2016.2579198

Shrestha, A., & Mahmood, A. (2019). Review of Deep Learning Algorithms and Architectures. *IEEE Access: Practical Innovations, Open Solutions, 7,* 53040–53065. doi:10.1109/ACCESS.2019.2912200

Sicari, S., Rizzardi, A., Grieco, L., & Coen-Porisini, A. (2015). Security, privacy and trust in internet of things: The road ahead. *Computer Networks, 76*(Suppl. C), 46–164. doi:10.1016/j.comnet.2014.11.008

SigFox. (n.d.). www.sigfox.com

Singh, S., Singh, K., & Saxena, A. (2020). Security Domain, Threats, Privacy issues in the Internet of Things (IoT): A Survey. *Proceedings of the Fourth International Conference on I-SMAC (IoT in Social, Mobile, Analytics, and Cloud) (I-SMAC),* 287–294.

Singh, H., Gupta, D. L., & Singh, A. K. (2014). Quantum key distribution protocols: A review. *Journal of Computational Engineering, 16*(2), 1–9. doi:10.1155/2014/785294

Solangi, Z. A., Solangi, Y. A., & Chandio, S., bin Hamzah, M. S., & Shah, A. (2018, May). The future of data privacy and security concerns in Internet of Things. In *2018 IEEE International Conference on Innovative Research and Development (ICIRD)* (pp. 1-4). IEEE.

Soumyalatha. (2016). Study of IoT: Understanding IoT architecture, applications, issues and challenges. *International Journal of Advanced Networking & Applications, 478.*

Souri, A., & Hosseini, R. (2018). A state-of-the-art survey of malware detection approaches using data mining techniques. *Human-centric Computing and Information Sciences, 8*(1), 1–22. doi:10.118613673-018-0125-x

Štitilis, D., Pakutinskas, P., Kinis, U., & Malinauskaitė, I. (2016). Concepts and principles of cyber security strategies. *Journal of Security & Sustainability Issues, 6*(2), 197–210. doi:10.9770/jssi.2016.6.2(1)

Stojmenovic, I., & Wen, S. (2014). The fog computing paradigm: scenarios and security issues. *Proceedings of the Federated Conference on Computer Science and Information Systems (FedCSIS' 14)*, 1–8. 10.15439/2014F503

Su, J., Vasconcellos, V. D., Prasad, S., Daniele, S., Feng, Y., & Sakurai, K. (2018). Lightweight classification of IoT malware based on image recognition. *Proc. IEEE 42nd Annu. Comput. Softw. Appl. Conf. (COMPSAC), 2*, 664–669.

Sudeendra Kumar, K., Sahoo, S., Mahapatra, A., Swain, A. K., & Mahapatra, K. K. (2018). Security enhancements to system on chip devices for IoT perception layer. *Proceedings - 2017 IEEE International Symposium on Nanoelectronic and Information Systems, INIS 2017*, 151–156. 10.1109/iNIS.2017.39

Swamy, S. N., & Kota, S. R. (2020). An empirical study on system level aspects of Internet of Things (IoT). *IEEE Access: Practical Innovations, Open Solutions, 8*, 188082–188134. doi:10.1109/ACCESS.2020.3029847

Syed, J. N., Shree, K. S., Shurjeel, W., Mohammad, N., & Md. Asaduzz, A. (2019). Quantum Machine Learning for 6G Communication Networks: State-of-the-Art and Vision for the Future. *IEEE Access: Practical Innovations, Open Solutions, 7*, 46317–46350. doi:10.1109/ACCESS.2019.2909490

Tank, B., Upadhyay, H., & Patel, H. (2016, August). Mitigation of privacy issues in IoT by modifying CoAP. In *2016 International Conference on Inventive Computation Technologies (ICICT)* (Vol. 3, pp. 1-4). IEEE.

Taylor, L., Fici, G. P., & Hersent, O. (2011). Interconnecting Zigbee & M2M Networks. *ETSI M2M Workshop*, 1-18.

Thai, M.T., Wang, F., & Du, D-Z. (2008). Coverage problems in wireless sensor networks: designs and analysis. *International Journal of Sensor Networks, 3*(3), 191–200.

Tian, Z., Sun, Y., Su, S., Li, M., Du, X., & Guizani, M. (2019). *Automated attack and defense framework for 5G security on physical and logical layers*. Available: https://arxiv.org/abs/1902.04009

TinyO. S. (2022). https://en.wikipedia.org/wiki/TinyOS

Tomita, A., Nambu, Y., & Tajima, A. (2005). Recent progress in quantum key transmission. *NEC J. of Adv. Tech, 2*(1), 84-91.

Tyagi, A. K., & Sreenath, N. (2021). Cyber physical systems: Analyses, challenges and possible solutions. *Internet of Things and Cyber-Physical Systems*.

Tyagi, A. K. (2016, March). Article: Cyber Physical Systems (CPSs) – Opportunities and Challenges for Improving Cyber Security. *International Journal of Computers and Applications*, *137*(14), 19–27. doi:10.5120/ijca2016908877

Tyagi, A. K., & Aghila, G. (2011). A wide scale survey on botnet. *International Journal of Computers and Applications*, *34*(9), 10–23.

Tysowski & Hasan. (n.d.). Hybrid attribute-and re-encryption based key management for secure and scalable mobile applications in clouds. *IEEE Trans. Cloud Comput.*, *1*(2), 172–186.

Vakhter, V., Soysal, B., Schaumont, P., & Guler, U. (2022). Threat Modeling and Risk Analysis for Miniaturized Wireless Biomedical Devices. *IEEE Internet of Things Journal*, 1. doi:10.1109/JIOT.2022.3144130

Vangelista, L., Zanella, A., & Zorzi, M. (2015). Long-range IoT technologies: The dawn of LoRaTM. In *Future Access Enablers of Ubiquitous and Intelligent Infrastructures* (pp. 51–58). Springer.

Vasileios, P.R., Sotirios, S., Panagiotis, S., Shaohua, W., George, K. K., & Sotirios K.G. (2021). Machine Learning in Beyond 5G/6G Networks—State-of-the-Art and Future Trends. *Electronics*, *10*, 2786. . doi:10.3390/electronics10222786

Wang, X., Xia, F., & Rodrigues, J. (2019). Joint Computation Offloading, Power Allocation, and Channel Assignment for 5G-Enabled Traffic Management Systems. *IEEE Transactions on Industrial Informatics*. . doi:10.1109/TII.2019.2892767

Wang, Y., Uehara, T., & Sasaki, R. (2015). Fog computing: Issues and challenges in security and forensics. *IEEE 39th Annual Computer Software and Applications Conference, 3*, 53–59.

Wang, D., Cheng, H., He, D., & Wang, P. (2018). On the challenges in designing identity-based privacy-preserving authentication schemes for mobile devices. *IEEE Systems Journal*, *12*(1), 916–925.

Wang, J., & Zhong, N. (2006). Efficient point coverage in wireless sensor networks. *Journal of Combinatorial Optimization*, *11*(3), 291–304. doi:10.100710878-006-7909-z

Warner, M. (2012). Cybersecurity: A pre-history. *Intelligence and National Security*, *27*(5), 781–799. doi:10.1080/02684527.2012.708530

Wazid, Das, Hussain, & Succi, & Rodrigues. (2019). Authentication in cloud-driven IoT-based big data environment: Survey and outlook. *Journal of Systems Architecture*, *97*, 185–196.

Wazid, M., Bagga, P., Das, A. K., Shetty, S., Rodrigues, J. J. P. C., & Park, Y. (2019). AKM-IoV: Authenticated key management protocol in fog computing based Internet of vehicles deployment. *IEEE Internet Things J.*, *6*(5), 8804–8817.

Wazid, M., & Das, A. K. (2016). An efficient hybrid anomaly detection scheme using K-means clustering for wireless sensor networks. *Wireless Personal Communications*, *90*(4), 1971–2000.

Wazid, M., & Das, A. K. (2017). A secure group-based blackhole node detection scheme for hierarchical wireless sensor networks. *Wireless Personal Communications*, *94*(3), 1165–1191.

Wazid, M., Das, A. K., Hussain, R., Succi, G., & Rodrigues, J. J. P. C. (2019). Authentication in cloud-driven IoT-based big data environment: Survey and outlook. *Journal of Systems Architecture*, *97*, 185–196.

Wazid, M., Das, A. K., Kumari, S., & Khan, M. K. (2016). Design of sinkhole node detection mechanism for hierarchical wireless sensor networks. *Security and Communication Networks*, *9*(17), 4596–4614.

Wazid, M., Das, A. K., Kumar, N., Conti, M., & Vasilakos, A. V. (2018). A novel authentication and key agreement scheme for implantable medical devices deployment. *IEEE Journal of Biomedical and Health Informatics*, *22*(4), 1299–1309.

Wazid, M., Das, A. K., Kumar, N., Vasilakos, A. V., & Rodrigues, J. J. P. C. (2019). Design and analysis of secure lightweight remote user authentication and key agreement scheme in Internet of drones deployment. *IEEE Internet Things J.*, *6*(2), 3572–3584.

Wazid, M., Das, A. K., & Lee, J.-H. (2019). User authentication in a tactile Internet based remote surgery environment: Security issues, challenges, and future research directions. *Pervasive and Mobile Computing*, *54*, 71–85.

Wazid, M., Das, A. K., Odelu, V., Kumar, N., & Susilo, W. (2020). Secure remote user authenticated key establishment protocol for smart home environment. *IEEE Transactions on Dependable and Secure Computing*, *17*(2), 391–406.

Wazid, M., Das, A. K., Shetty, S., Gope, P., & Rodrigues, J. J. P. C. (2020). Security in 5G-Enabled Internet of Things Communication: Issues, Challenges and Future Research Roadmap. *IEEE Access: Practical Innovations, Open Solutions*, *8*, 1–25. https://doi.org/10.1109/ACCESS.2020.3047895

Wazid, M., Das, A. K., & Vasilakos, A. V. (2018). Authenticated key management protocol for cloud-assisted body area sensor networks. *Journal of Network and Computer Applications*, *123*, 112–126.

Wazid, M., Dsouza, P. R., Das, A. K., Bhat, V. K., Kumar, N., & Rodrigues, J. J. P. C. (2019). RAD-EI: A routing attack detection scheme for edge-based Internet of Things environment. *International Journal of Communication Systems*, *32*(15), 4024. doi:10.1002/dac.4024

Weis, R., & Lucks, S. (2000). The performance of modern block ciphers in java. In J.-J. Quisquater & B. Schneier (Eds.), *Smart card research and applications*. Springer.

Whitmore, A., Agarwal, A., & Xu, L. D. (2015). The Internet of Things—A survey of topics and trends. *Information Systems Frontiers*, *12*(2), 261-274.

Worner, D., & von Bomhard, T. (2014). When your sensor earns money: Exchanging data for cash with bitcoin. *Proceedings of the 2014 ACM International Joint Conference on Pervasive and Ubiquitous Computing: Adjunct Publication*, 295-298.

Wu, B., Xu, K., Li, Q., Liu, Z., Hu, Y.-C., Zhang, Z., Du, X., Liu, B., & Ren, S. (2019). Decentralized and automated incentives for distributed IoT system detection. *Proc. IEEE 39th Int. Conf. Distrib. Comput. Syst. (ICDCS)*, 1106–1116.

Wu, L., Wang, J., Choo, K.-K.-R., & He, D. (2019). Secure key agreement and key protection for mobile device user authentication. *IEEE Transactions on Information Forensics and Security*, *14*(2), 319–330.

Xu & Sahni. (2006). *Approximation Algorithms For Wireless Sensor Deployment*. Academic Press.

Xu, X., Chen, Y., Zhang, X., Liu, Q., Liu, X., & Qi, L. (2019). A blockchain-based computation offloading method for edge computing in 5G networks. *Software, Practice & Experience*.

Yang, S., Wu, Y., & Xuan, J. (2007). *Time Series Analysis in Engineering Application*. HUST Press.

Yang, Y., Wu, L., Yin, G., Li, L., & Zhao, H. (2017). A survey on security and privacy issues in Internet-of-Things. *IEEE Internet Things J.*, *4*(5), 1250–1258.

Ye, D., & Zhang, M. (2017). A self-adaptive sleep/wake-up scheduling approach for wireless sensor networks. *IEEE Transactions on Cybernetics*, *48*(3), 979–992. doi:10.1109/TCYB.2017.2669996 PMID:28278488

Yi, S., Hao, Z., Qin, Z., & Li, Q. (2016). Fog computing: Platform and applications. *Proceedings - 3rd Workshop on Hot Topics in Web Systems and Technologies, HotWeb 2015*, 73–78. 10.1109/HotWeb.2015.22

Yi, S., Qin, Z., & Li, Q. (2015). Security and privacy issues of fog computing: A survey. *10th International Conference on Wireless Algorithms, Systems, and Applications*, 1–10.

Yick, J., Mukherjee, B., & Ghoshal, D. (2008). Wireless Sensor Networks Survey. *Computer Networks*, *52*, 2292–2330.

Yogesh, K., Komalpreet, K., & Gurpreet, S. (2020). Machine Learning Aspects and its Applications Towards Different Research Areas. *IEEE International Conference on Computation, Automation and Knowledge Management (ICCAKM)*. 10.1109/ICCAKM46823.2020.9051502

Yousefpour, A., Ishigaki, G., Gour, R., & Jue, J. P. (2018). On Reducing IoT Service Delay via Fog Offloading. *IEEE Internet of Things Journal*, *5*(2), 998–1010. doi:10.1109/JIOT.2017.2788802

Yu, J., Ren, K., Wang, C., & Varadharajan, V. (2015). Enabling cloud storage auditing with key-exposure resistance. *IEEE Transactions on Information Forensics and Security*, *10*(6), 1167–1179.

Zhang, X., Cao, Z., & Dong, W. (2020). Overview of Edge Computing in the Agricultural Internet of Things: Key Technologies, Applications, Challenges. *IEEE Access: Practical Innovations, Open Solutions*, *8*, 141748–141761. doi:10.1109/ACCESS.2020.3013005

Zhang, Y., & Wen, J. (2015). An IoT electric business model based on the protocol of bitcoin. *18th International Conference on Intelligence in Next Generation Network*, 184-191.

Zhang, Z. K., Cho, M. C. Y., Wang, C. W., Hsu, C. W., Chen, C. K., & Shieh, S. (2014). IoT security: Ongoing challenges and research opportunities. *7th IEEE International Conference on Service-Oriented Computer Applications*, 230–234.

Zhao, Y., Zhai, W., Zhao, J., Zhang, T., Sun, S., Niyato, D., & Yan Lam, K. (2020). *A Survey of 6G Wireless Communications: Emerging Technologies.* arXiv:2004.08549v3.

Zhao, Y., Zhai, W., Zhao, J., Zhang, T., Sun, S., Niyato, D., & Yan Lam, K. (2021). *A Comprehensive Survey of 6G Wireless Communications.* arXiv:2101.03889v2.

Zhou, J., Cao, Z., Dong, X., & Vasilakos, A. V. (2017). Security and privacy for cloud-based IoT: Challenges. *IEEE Communications Management, 55*(1), 26–33.

Zhou, W., Jia, Y., Peng, A., Zhang, Y., & Liu, P. (2019). The effect of IoT new features on security and privacy: New threats, existing solutions, and challenges yet to be solved. *IEEE Internet Things J., 6*(2), 1606–1616.

Zhou, Z., Das, S., & Gupta, H. (2004). Connected k-coverage problem in sensor networks. *Proc. of International Conference on Computer Communications and Networks (ICCCN'04)*, 373–378.

Zhu, S., Setia, S., & Jajodia, S. (2006). LEAP+: Efficient security mechanisms for large-scale distributed sensor networks. *ACM Transactions on Sensor Networks, 2*(4), 500–528.

Zyskind, G., Nathan, O., & Pentland, A. (2015). *Enigma: Decentralized computation platform with guaranteed privacy.* Academic Press.

About the Contributors

Biswa Mohan Sahoo is a Senior IEEE member and received his B.Tech and M.Tech degree in computer science and engineering from Biju Patnaik University of Technology, Odisha, India and PhD degree in computer science and engineering from Indian Institute of Technology (ISM), Dhanbad, India. He is currently working as an Assistant Professor at Manipal University Jaipur, India. He has published over seventy articles in peer-reviewed journals and conferences of international repute like Elsevier, Springer, IEEE, IOS Press, Taylor & Francis, etc. Apart from that, 07 Book Chapters, 2 Books and three international Patents under his credit. His current research interests include IoT, Wireless Sensor Networks, Swarm Intelligence, AI, and Image Processing. His currently focus on artificial intelligence approaches on sensor networks. His research interest area is wireless sensor network with Evolutionary Algorithm and Artificial intelligence. He serves as a reviewer of many peer-reviewed journals such as IEEE Journal, Wireless Personal Communication, Complex and Intelligent System, etc. He has also served as Technical Program Committee Member of several conferences of international repute.

Suman Avdhesh Yadav is a Senior IEEE member and received her B.Tech and M.Tech degrees in Software Engineering from Dr. A.P.J. Abdul Kalam Technical University, Uttar Pradesh, India. She is pursuing her Ph.D. degree in Computer Science and Engineering from Galgotias University, Uttar Pradesh, India. She is currently working as an Assistant Professor at Amity University, Uttar Pradesh, India. She has published over twelve articles in peer-reviewed journals and conferences of international repute like Elsevier, Springer, IEEE, Taylor & Francis, etc. Apart from that, 02 Book Chapters and 02 Indian Patents are under her credit. Her current research interests include IoT, Wireless Sensor Networks, Network Security, and AI. Her current focus in WSN is improving the lifetime and efficiency of sensor networks. She serves as a reviewer of many peer-reviewed journals and conferences. She has served as Conference Secretary of several IEEE conferences like ICTAI2021, ICIPTM2022, ICIEM2022, and ICCAKM2022. She has also served as a Technical Program Committee Member of several conferences of international repute.

* * *

Pooja Chaturvedi is currently working as Assistant Professor in Department of Computer Science and Engineering in Institute of Technology, Nirma University, Ahmedabad, Gujrat.She has 6 years of teaching experience. She has completed Ph. D., M. Tech. and B. Tech. in Computer Science and Engineering in the years 2018. 2012 and 2010 respectively. She has published more that 30 paper in various referred journals and conferences. She has also served as a Reviewer for the referred journals and international conferences. Her research areas includes Wireless Sensor Networks and Artificial Intelligence.

Khushi Dadhich is currently pursuing B.Tech in the department of Computer Science from Amity University, Greater Noida. Her current research interests include cloud services and the application of IoT, AI & ML and special focus on the application of IoT. In addition to her academics she has also participated in hackathons and different technology based events.

Ajai Daniel is currently working as Professor and H. O. D. in the Department of Computer Science and Engineering, Madan Mohan Malaviya University of Technology, Gorakhpur. He has 36 years of teaching experience and has guided a number of research scholars. He has published more than 130 papers in international journals and conferences. His research areas include Adhoc network, artificial intelligence, and protocols.

Lipsa Das is Assistant Professor at the department of Electronics & Communication Engineering at Amity University with 10 years of teaching experience both online and offline. She received B'Tech from BPUT, Rourkela and received M'Tech from IIT Kharagpur. Currently, she is pursuing her Ph.D. degree from Amity University, Noida. Her research is situated in the field of Technology & Innovation, with a special focus on VLSI Design, cloud services and the application of IoT, AI & ML. She has published several research papers in international conferences and also published no of Indian patents. In addition to her teaching experience, she has accomplished many administrative activities.

Sumit Dhariwal has a Bachelor of Engineering degree and Master of Technology degree. He obtained a BE degree in Rajiv Gandhi Technical University's Computer Science and Engineering Branch in 2008 and he obtained an MTech degree. Samrat Ashok of Technological Institute's Information & Technology Branch in 2011. Since 2018 he is now the Research Scholar of the Malaysia University of Science and Technology of the Department of Information and Technology in Malaysia. He worked as Assistant Professor in Technical Universities. And the author has also contributed a lot in the conference of the International Conference and Conference on the National & International Journal Journals.

Tukkappa K. Gundoor obtained the MCA. degree in Computer Science from the Visvesvaraya Technological University in 2018. he is currently working as a Research Scholar under the guidance of Dr. Sridevi, Professor, Department of Computer Science, Karnatak University, Dharwad. Received research fellowship from DST, KSTePS), Govt. of Karnataka. Area of research is "Network Security." At present working on the research topic: "Study and design of effective algorithm to detect Non-Portable malicious files". IEEE student member (Member no: 94567049), IAENG (International Association of Engineers) (Member no: 291948) and DRDO (Defence Research and Development Organisation) diat certified artificial intelligence professional.

Sridhar Iyer received the B.E. degree in Electronics and Telecommunications Engineering from Mumbai University, India in 2005. He received the M.S. degree in Electrical and Communication Engineering from New Mexico State University, U.S.A. in 2008, and the Ph.D. degree from Delhi University, India in 2017. Currently, he an Associate Professor in the Department of ECE, KLE Technological University, MSSCET-Belagavi Campus, India. His research interests include the architectural, algorithmic, and performance aspects of the optical networks, with current emphasis on efficient design and resource optimization in the space division multiplexing enabled flexi-grid Elastic optical networks, and spectrum allocation and management issues in 6G Communications.

Varsha Jayaprakash is a final year student of B.Tech at Vellore Institute of Technology, Chennai.

Manojkumar T. Kamble obtained his M.Sc. degree in Computer Science from the Rani Channamma University, Belagavi in 2018. Currently working as a Research Scholar under the guidance of Dr Sridevi, Professor, Department of Computer Science, Karnatak University, Dharwad. Received research fellowship from DST, KSTePS), Govt. of Karnataka. The area of research is "Network Security ". At present working on the research topic: "Study and design of effective malware detection approach using memory forensics". IEEE student member (Member ID: 95536730), IAENG (International Association of Engineers) (Member ID: 214266) and DRDO (Defence Research and Development Organisation) DIAT certified cyber security participated.

Kamalendu Pal is with the Department of Computer Science, School of Mathematics, Computer Science and Engineering, City, University of London. Kamalendu received his BSc (Hons) degree in Physics from Calcutta University, India, Postgraduate Diploma in Computer Science from Pune, India, MSc degree in Software Systems Technology from the University of Sheffield, Postgraduate Diploma in Artificial Intelligence from Kingston University, MPhil degree in Computer Science from University College London, and MBA degree from the University of Hull, United Kingdom. He has published over seventy research articles (including book chapters) in the scientific community with research papers in the ACM SIGMIS Database, Expert Systems with Applications, Decision Support Systems, and conferences. His research interests include knowledge-based systems, decision support systems, blockchain technology, software engineering, and service-oriented computing. He is on the editorial board in an international computer science journal. Also, He is a member of the British Computer Society, the Institution of Engineering and Technology, and the IEEE Computer Society.

Rahul Jashvantbhai Pandya completed his M.Tech. from the Electrical Engineering Department, Indian Institute of Technology, Delhi, New Delhi, in 2010. He completed his Ph.D. from Bharti School of Telecommunication and Management, Indian Institute of Technology, Delhi, in 2014. He worked as the Sr. Network Design Engineer in the Optical Networking Industry, Infinera Pvt. Ltd., Bangalore, from 2014 to 2018. Later, from 2018 to 2020, he worked as the Assistant Professor, ECE Department, National Institute of Technology, Warangal. Currently, he is working Electrical Engineering department, Indian Institute of Technology, Dharwad. His research areas are Wireless Communication, Optical Communication, Optical Networks, Computer Networks, Machine Learning, and Artificial Intelligence.

Apoorva Shripad Patil obtained the M.Sc. degree in Computer Science from the Karnatak University, Dharwad in 2019. She is currently working as a Research Scholar under the guidance of Dr. Sridevi, Professor, Department of Computer Science, Karnatak University, Dharwad. Received the Research Fellowship from DST (Department of Science and Technology), Karnataka Science and Technology Promotion Society (KSTePS), Govt. of Karnataka. Area of research is "Network Security ". At present working on the research topic: "Edge Computing for IoT: Some Issues and Approaches". IEEE Student member (member id 98456212). Cyber Security Course by Defence Institute of Advanced Technology(DIAT), DRDO. (Diat Certified Information Assurance Intermediate Level).

Sridevi, Professor, Department of Computer Science, Karnatak University, Dharwad, obtained Ph.D. from Mangalore University, Mangalore, in 2017. Research interest includes Advanced Computer Networks, Mobile and Wireless Communication, Network security, Cloud Computing and Internet of Things. At present guiding four research scholars and One M.Phil. awarded. And also, Reviewer for the Journal of Advances in Mathematics and Computer Science. Published more than 30 research papers in national and international journal.

Srividhya S. R. has graduated from SCSVMV University with Bachelor's Degree in Information Technology. she has received her M.E. degree in Computer Science and Engineering from Anna University, Chennai in 2015. She has more than five years of teaching experience in the Department of Computer Science and Engineering in various colleges. She is currently serving as Assistant Professor at the School of Computing, Sathyabama Institute of Science and Technology, Chennai. She is currently pursuing her research work in medical image processing and deep learning. Her research interests are AI, image processing, and Big Data. She has presented her research findings in various reputed national/international journals and also in the proceedings of national/ international conferences.

Avani Sharma completed PhD from NIT, Rajasthan and M.Tech from LNMIIT. She research area include Internet of Things, Soft Security for IoT, MANET Security and Machine Learning. She has published research articles in various international conferences and journals. She is currently working as an Assistant Professor in Manipal University Jaipur, India.

Smita Sharma is currently working as assistant professor at Amity University and leading the ECE/EEE departments. She has completed her graduation degree in Electronics Communication from Galgotias college of engineering and technology and done her M.Tech from Madan Mohan Malviya engineering college, Gorakhpur. Currently, she is pursuing her Ph.D. degree from Uttrakhand Technical University. She has published several research papers in international conferences and journals. She is currently guiding graduate students with various projects related to wireless sensor networks, and computer networks, and graduate students with various projects related to wireless sensor networks, computer networks, and embedded systems. She has vast experience in research and teaching and has contributed experience in the field of research and teaching and has contributed to the development of the various centers of excellence at Amity University. With 10 years of teaching experience have managed various administrative and educational activities online and offline.

Amit Kumar Tyagi (GATE, NPwD-JRF, UGC-NET, and ICAR-NET) received his Ph.D degree in 2018 from Pondicherry Central University, Puducherry, India, in area of "Vehicular Ad-hoc Networks". His research interests include Formal Language Theory, Smart and Secure Computing, Privacy (including Genomic Privacy), Machine Learning with Big data, Blockchain Technology, Cyber Physical System etc. He has completed his M.Tech degree from the Pondicherry Central University, Puducherry, India. He joined the Lord Krishna College of Engineering, Ghaziabad (LKCE) for the periods of 2009-2010, and 2012-2013. With more than 08 (Eight) years of teaching and research experience across India, currently he is working as an Assistant Professor in Vellore Institute of Technology, Chennai Campus, Chennai 600127, Tamilnadu, India. Additionally, He is also the recipient/ awarded of the GATE and NPwD-JRF fellowship in 2009, 2016 and 2013. He has been published one major book titled "Know Your Technical (IT) Skills". Also He is a member of various Computer/ Research Communities like IEEE, ISOC, CSI, ISTE, DataScience, MIRLab, etc.

Index

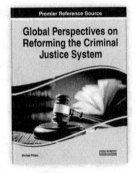

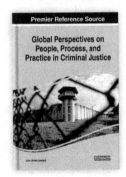

Ensure Quality Research is Introduced to the Academic Community

Become an Evaluator for IGI Global Authored Book Projects

Premier Reference Source

Stabilizing Currency and Preserving Economic Sovereignty Using the Grondona System

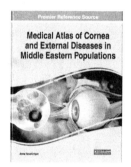

Premier Reference Source

Medical Atlas of Cornea and External Diseases in Middle Eastern Populations

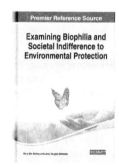

Premier Reference Source

Examining Biophilia and Societal Indifference to Environmental Protection

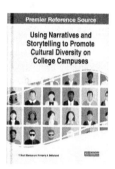

Premier Reference Source

Using Narratives and Storytelling to Promote Cultural Diversity on College Campuses

The overall success of an authored book project is dependent on quality and timely manuscript evaluations.

Applications and Inquiries may be sent to:
development@igi-global.com

Applicants must have a doctorate (or equivalent degree) as well as publishing, research, and reviewing experience. Authored Book Evaluators are appointed for one-year terms and are expected to complete at least three evaluations per term. Upon successful completion of this term, evaluators can be considered for an additional term.

If you have a colleague that may be interested in this opportunity, we encourage you to share this information with them.

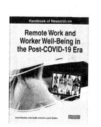

Printed in the United States
by Baker & Taylor Publisher Services